DEFINITION OF PAIN AND DISTRESS AND REPORTING REQUIREMENTS FOR LABORATORY ANIMALS

PROCEEDINGS OF THE WORKSHOP
HELD JUNE 22, 2000

Committee on Regulatory Issues in Animal Care and Use
Institute for Laboratory Animal Research
National Research Council

NATIONAL ACADEMY PRESS
Washington, D.C.

NATIONAL ACADEMY PRESS   **2101 Constitution Avenue, NW**   **Washington, DC 20418**

NOTICE: The project that is the subject of these proceedings was approved by the Governing Board of the National Research Council, whose members are drawn from the councils of the National Academy of Sciences, the National Academy of Engineering, and the Institute of Medicine. The members of the committee responsible for the project were chosen for their special competences and with regard for appropriate balance.

This study was supported by Grant No. N01-0D-4-2139 between the National Academies and the National Institutes of Health of the U.S. Department of Health and Human Services. Any opinions, findings, conclusions, or recommendations expressed in this publication are those of the author(s) and do not necessarily reflect the views of the organizations or agencies that provided support for the project.

International Standard Book Number 0-309-07291-3

Definition of Pain and Distress and Reporting Requirements for Laboratory Animals: is available from the National Academy Press, 2101 Constitution Ave., NW, Washington, DC 20418; 800-624-6242 or 202-334-3313 in the Washington metropolitan area; Internet: www.nap.edu.

THE NATIONAL ACADEMIES

National Academy of Sciences
National Academy of Engineering
Institute of Medicine
National Research Council

The **National Academy of Sciences** is a private, nonprofit, self-perpetuating society of distinguished scholars engaged in scientific and engineering research, dedicated to the furtherance of science and technology and to their use for the general welfare. Upon the authority of the charter granted to it by the Congress in 1863, the Academy has a mandate that requires it to advise the federal government on scientific and technical matters. Dr. Bruce M. Alberts is president of the National Academy of Sciences.

The **National Academy of Engineering** was established in 1964, under the charter of the National Academy of Sciences, as a parallel organization of outstanding engineers. It is autonomous in its administration and in the selection of its members, sharing with the National Academy of Sciences the responsibility for advising the federal government. The National Academy of Engineering also sponsors engineering programs aimed at meeting national needs, encourages education and research, and recognizes the superior achievements of engineers. Dr. William A. Wulf is president of the National Academy of Engineering.

The **Institute of Medicine** was established in 1970 by the National Academy of Sciences to secure the services of eminent members of appropriate professions in the examination of policy matters pertaining to the health of the public. The Institute acts under the responsibility given to the National Academy of Sciences by its congressional charter to be an adviser to the federal government and, upon its own initiative, to identify issues of medical care, research, and education. Dr. Kenneth I. Shine is president of the Institute of Medicine.

The **National Research Council** was organized by the National Academy of Sciences in 1916 to associate the broad community of science and technology with the Academy's purposes of furthering knowledge and advising the federal government. Functioning in accordance with general policies determined by the Academy, the Council has become the principal operating agency of both the National Academy of Sciences and the National Academy of Engineering in providing services to the government, the public, and the scientific and engineering communities. The Council is administered jointly by both Academies and the Institute of Medicine. Dr. Bruce M. Alberts and Dr. William A. Wulf are chairman and vice chairman, respectively, of the National Research Council.

COMMITTEE ON REGULATORY ISSUES IN ANIMAL CARE AND USE

Adele Douglass, American Humane Association, Washington, D.C.
Randall J. Nelson, Department of Anatomy and Neurobiology, University of
Tennessee, Memphis, Tenn.
William S. Stokes, Environmental Toxicology Program, National Institute of
Environmental Health Science, Research Triangle Park, N.C.
Jerrold Tannenbaum, Department of Population and Health Reproduction,
University of California School of Veterinary Medicine, Davis, Calif.
Joanne Zurlo, Center for Alternatives to Animal Testing, Johns Hopkins
School of Hygiene and Public Health, Baltimore, Md.

Staff

Ralph B. Dell, Director
Kathleen A. Beil, Administrative Assistant
Susan S. Vaupel, Editor
Marsha K. Williams, Project Assistant

Shirley M. Tilghman, Department of Molecular Biology, Princeton University, Princeton, N.J.

Raymond L. White, Department of Oncological Sciences, University of Utah School of Medicine, Salt Lake City, Utah

Staff

Warren Muir

Preface

In this first in a proposed series of workshops on regulatory issues in animal care and use, the Institute for Laboratory Animal Research (ILAR) has addressed the existing and proposed requirements for reporting pain and distress in laboratory animals. The Animal Welfare Act, administered by the Animal and Plant Health Inspection Service of the United States Department of Agriculture (USDA), mandates that pain and distress in laboratory animals be minimized. USDA is considering two policy changes with regard to this specific mandate. Firstly, since there has been no functional definition of "distress," USDA has prepared such a definition and requested feedback from the scientific community on its usefulness for regulatory and reporting requirements. (See Appendix B.)

The second issue concerns the pain and distress categorization scheme for reporting to USDA. Various groups and individuals have questioned the efficacy of the current categories, and specific changes have been proposed by the Humane Society of the United States. USDA is considering these and other potential changes to the existing scheme. Thus, given these potential changes to animal welfare policy, the aim of the ILAR/NIH joint workshop was to provide feedback to the USDA. The speakers were asked to address these two issues as well as to comment upon whether the information contained in the 1992 ILAR report *Recognition and Alleviation of Pain and Distress in Laboratory Animals* is still useful to investigators in assisting them to comply with regulations. The speakers provided perspectives based on their individual expertise in the areas of science of pain and distress, animal welfare policy, protocol review, and/or as representatives of relevant organizations or institutions. The following proceedings are an edited transcript of their presentations.

Contents

Introduction

Ralph B. Dell

Director
Institute for Laboratory Animal Research (ILAR)

Welcome. The purpose of this workshop is to talk to representatives of the USDA about principles and definitions pertaining to the recognition and alleviation of pain and distress in laboratory animals. Several questions are related:

- Can we compose a definition?
- Can we produce language that will inform both the people who will carry out the regulations and the people who will inspect the process to determine whether, indeed, the institution is responding to the regulations appropriately?

An underlying question is:

- Can we come up with language or words that go across all species or should we choose language that is somewhat dependent on the species we are describing.

This workshop provides an opportunity for the speakers and members of the audience to engage in a discussion of the definitions of pain, distress, and how one can recognize and alleviate the pain and distress that can occur in the course of using animals in biomedical research as well as in education and testing. The purpose of the workshop is to focus on the proposed wording for the USDA to use in writing regulations that will implement the Animal Welfare Act. Because the Animal Welfare Act contains the phrase "pain and distress," the USDA must define those terms to implement the act.

I want to thank all of you for coming and participating in this important event. We have organized the program to have presentations by people on the regulatory side first and then presentations by people who are students of animal welfare/animal behavior as well as pain physiology. The latter part of the program will involve people who will build on the foregoing presentation and will propose language and definitions for possible use in policies and in regulations.

In 1992, ILAR published a report titled *Recognition and Alleviation of Pain and Distress in Laboratory Animals* (NRC 1992). Other questions we should address today are the following:

- Are the principles and definitions articulated in that report still current?
- Can they be used by practitioners and by researchers to evaluate and treat pain and distress in the laboratory animals they are using?
- Are the principles and the language in that report clear (unambiguous) enough for the USDA to use in their policies and regulations?

If not, then one of the tasks of ILAR will be to set up a committee to revise that report for consistency with modern and current thinking. That matter is another aspect of today's workshop.

I would now like to introduce Dr. Ron DeHaven, who is well known to everyone in the room. Dr. DeHaven is Deputy Administrator of Animal Care for USDA/APHIS.

REFERENCE

NRC [National Research Council]. 1992. Recognition and Alleviation of Pain and Distress in Laboratory Animals. Washington, D.C.: National Academy Press.

Pain and Distress: USDA Perspective

W. Ron DeHaven

Deputy Administrator, Animal Care
USDA/APHIS
Washington, D.C.

I want to extend my thanks to ILAR (especially Dr. Ralph Dell) and to OLAW, NIII, for cosponsoring this meeting. Clearly, pain and distress is an issue looming very large on the horizon. It has been smoldering for approximately 5 years, but I suspect it will become very active in the next 3 to 5 years. USDA is here to listen because we will be dealing with this issue very actively in the immediate future. We will need to deal with the following aspects of the issue:

1. Because the focus in the past clearly has been on pain and far less on distress, we need to develop a definition of distress. The definition must work both for the biomedical research community and from a regulator's standpoint.

2. Alternatives to painful and distressful procedures must be considered, as required by the Animal Welfare Act and, of course, in the regulations that implement the Act.

3. To require minimization of pain and distress presupposes that one can recognize pain and distress in animals when it occurs.

4. There is much discussion now about our current pain and distress categorization system and whether it should be revised.

We cannot talk about any of these things without also considering the potential impact if the USDA should decide to regulate rats, mice, and/or birds. The burden that would be imposed would not be confined to regulating those species, but also would involve minimizing pain and distress in those animals and imposing reporting requirements because they represent 90% or more of the animals being used in biomedical research.

POLICY 12

We have just revised Policy 12 (Table 1 and Appendix A) after a considerable amount of discussion and controversy. This policy, which requires consideration of alternatives to painful and distressful procedures, states that the database search remains the most effective and efficient method of satisfying the requirement of considering alternatives to painful and distressful procedures. At the same time, it recognizes that this method might not always be the best, or it might be one of multiple methods, used to satisfy the requirement. We believe the test should be applied by the IACUC as well as our inspectors and is two-fold. The first part of the test involves whether there is enough information for the IACUC to make a determination as to whether the principal investigator has made a good faith and reasonable effort to determine what alternatives are available. The second part of the test focuses on whether adequate consideration was given to bona fide alternatives that were identified.

POLICY 11

Policy 11 relates to the minimization of pain and distress and reporting the same. We are in the process of revising that document. Clearly, whatever we produce is going to focus on minimizing pain and distress to the extent practical, as does the AWA and the regulations, and it will also include a definition of distress.

The draft version of this policy (Table 2 and Appendix A) recognizes that there might be other methods of relief besides anesthetics, analgesics, and tranquilizers. For example, there may be therapeutic agents (e.g., antibiotics) that in a disease process would provide some relief from pain and distress. In addition, other chemical agents or methods may provide relief from pain and distress.

The new policy will recognize that any given procedure (complete Freund's adjuvant might be an example) can be appropriately categorized in our current system under column C, D, or E, depending on the circumstances of how it is employed. On a retrospective evaluation of each animal on a case-by-case basis, it is certainly possible for some procedures to be categorized appropriately in any of the three current categories. Prospective reporting is acceptable, and retrospective

TABLE 1 Policy 12: Consideration of Alternatives

- Database search remains the most effective and efficient method.
- Other sources may be considered.
- Sufficient information is provided for the IACUC to determine whether a good faith and reasonable effort was made.
- Adequate consideration is given to bona fide alternatives found.

TABLE 2 Policy 11: Minimization of Pain and Distress—Draft

- Defines distress
- Recognizes other methods of relief—therapeutic agents, specialized nursing care, behavioral conditioning
- Recognizes that a single procedure may be placed in column C, D, or E, depending on circumstances
- Explains prospective versus retrospective reporting
- Provides column descriptions and classification examples

---

reporting will clearly be allowed. As we currently have it drafted, this new Policy 11 would include descriptions of the three columns as found in our annual report. It would provide examples of what types of procedures would be categorized appropriately in each of those columns.

This definition of painful procedure is direct from the regulations: "Any procedure that would reasonably be expected to cause more than slight or momentary pain or distress in a human being to which the procedure was applied, that is, pain in excess of that caused by injections or other minor procedures." The definition of distress in the current version of the rewrite of Policy 11 is as follows: "A state in which an animal cannot escape from or adapt to the external or internal stressors or conditions that it experiences resulting in negative effects upon its well-being." We certainly welcome your comments on this working definition of distress. This definition of distress is for AWA purposes. No doubt there are other scenarios and situations where this definition of distress might not be appropriate.

We believe that the definition of distress for AWA purposes must be in excess of that which an animal would experience by routine husbandry or handling practices. In other words, simply caging an animal in a laboratory would not be considered distress in the context of the AWA and our regulations. Additionally, to distinguish stress from distress is critical in that the requirements for minimization of pain and distress apply to "distress" and not to "stress."

We believe adherence to Principle IV of the US Government Principles is important and have included it in our working draft of the policy. It states that unless the contrary is established, investigators should consider that procedures that cause pain or distress in human beings will cause pain or distress in animals.

I mentioned earlier the concept of prospective versus retrospective reporting. Prospective reporting of pain or distress involves classifying animals in categories based on anticipation or expectation that the animals are likely to experience pain or distress and whether one plans to use any of the pain-relieving medications. In contrast, retrospective reporting is based on the actual observed presence or absence of pain or distress and the actual use or nonuse of pain-relieving measures.

EXPLANATION OF CATEGORIES

The descriptions we may include in a policy will not necessarily mimic or restate what the regulations state because the purpose of a policy is to provide clarification of the existing regulation. We hope that the following explanations and examples (among many) of the various categories will clarify these categories.

Column C

Description: Testing, teaching, or experiments involving no pain, distress, or use of pain-relieving medications. Again, recognize that pain or distress must be more than slight or momentary to be, by definition, a painful procedure.

Example: Collection of a blood sample from a peripheral vein not resulting in pain or distress. The animal remains calm throughout.

Column D

Description: Testing, teaching, or experiments involving pain or distress for which appropriate analgesic, anesthetic, or tranquilizing drugs are used.

Examples: (1) Use of an anesthetic agent to prevent pain or distress associated with intercardiac blood collection or perhaps any number of other types of procedures. (2) Use of a therapeutic agent to remedy an intentionally induced disease process. It would be a bit of a leap from where we are now, since current regulations recognize only anesthetics, analgesics, and tranquilizers or other pain-relieving medications as a means to appropriately classify procedures in Column D.

Column E

Description: Teaching, testing, or experiments involving pain or distress for which the use of appropriate anesthetic, analgesics, or tranquilizing drugs would have adversely affected the procedure's results or the interpretation.

Examples: (1) Collection of an intercardiac blood sample without the use of pain-relieving medications. (2) Presence of an experimentally induced disease process that causes pain or distress that is not relieved by the use of pain-relieving medications.

DISCUSSION

As I mentioned, there is a much discussion about our current pain and distress categorization scheme and whether it accomplishes what we want to accomplish. Clearly, there is a diversity of opinion on what we are trying to accomplish, and I suspect we will hear some of those opinions later today.

TABLE 3 Gaps in the Current System

• The P/D[a] classifications reported are a mixture of prospective *expectations* of pain and distress and retrospective *observed* pain and distress.
• Pain and distress are not differentiated when classifying or reporting.
• Pain- and distress-relieving measures other than drugs are not recognized in the regulations.
• Effectiveness of P/D relief is not addressed.
• Timeliness of P/D relief is not addressed.
• Intensity and duration of P/D are not addressed.
• No category exists for animals not benefiting from pain-relieving medications but for which there was no prohibition to their use.

[a]P/D, pain and/or distress.

Listed in Table 3 are some of the gaps or areas where we think our current system may not be adequate and therefore may provide a reason for changing those categories. First, the pain and distress classifications reported are a mixture of prospective expectations, or prospective reporting of pain and distress, and retrospective reporting based on actual observations. Maybe this is as it should be, a mixture; or maybe it should all be retrospective reporting. This is one of the areas where we need to have some dialogue.

The current system does not differentiate between pain and distress. Maybe it is not necessary that we do; maybe we should. Again, I think this debate is an area for dialogue.

It does not officially recognize pain-relieving measures other than the use of anesthetics, analgesics, or tranquilizers. Clearly, other methods, whether they are chemical or nonchemical, have the potential to reduce or alleviate pain or distress.

The effectiveness of the pain and distress relief that is utilized is not addressed. So, to provide a ridiculous example, one could perform thoracic surgery on an animal, provide postoperative analgesia consisting of an aspirin, and perhaps appropriately categorize that in Column D. I think that is subject to debate, but there is nothing in the regulation that requires any evaluation of whether a particular pain and distress relief was effective.

Timeliness is another issue. Is the pain-relieving medication given as soon as, or before, the animal actually perceives pain and distress, or does it occur afterward? If so, how long afterward?

The current system does not address intensity of pain or the duration of the pain or distress. I think it is important to recognize that an animal experiencing moderate pain for a prolonged period of time may need to be categorized differently from an animal experiencing intense pain but only for 1 or 2 seconds.

The final gap in the current system is slightly more difficult to understand. This complex issue has only recently come to light. Specifically, there is currently no accurate category in which to place an animal that does not benefit from the use of pain-relieving medications but for which there was no prohibition to the use of pain-relieving medications. Column E is for procedures for which pain-relieving medications could not be used because it is expected that the use of those medications would result in inaccurate data or inaccurate interpretations of the data. This description is in contrast to the animal that may have experienced pain or distress, was not provided pain-relieving medication, but for which there was no prohibition to the use of those medications.

As I said at the beginning, our primary reason for being here today is to listen. We will also be publishing a Federal Register Notice to gather input from the public and all interested parties with regard to this whole process. We hope that you will voice your opinion and, in doing so, will propose a solution and justification for any solution that you might suggest.

When we publish this Notice in the Federal Register, which we hope will be very soon, we hope that you will comment. We are anticipating a 60-day comment period for the definition of distress as well as the concept of changing our current pain and distress categorization scheme.

There are several ways to contact us through the Federal Register Notice, which will include specific instructions. In addition, you may submit your comments: via email (ace@usda.gov), telephone (301-734-4981), or fax (301-734-4978). Otherwise, the several USDA folks in this audience will be listening in earnest to the discussions throughout the day.

Pain, Distress, and Reporting Requirements: PHS Policy Perspective

Nelson Garnett

Director, Office of Laboratory Animal Welfare
National Institutes of Health
Rockville, Md.

It is my pleasure to be here with you this morning to talk about some very important public policy issues that are before us. Pain and distress (and especially distress, as we have heard) have been topics of several recent meetings, which have all contributed greatly to our understanding of these issues.

NIH is a partial sponsor of this workshop for several reasons. One reason is that Congress has raised a concern, which NIH has addressed through the commissioning of a study on regulatory burden. The study is associated with a variety of federal requirements, including not only animal welfare but also human subjects and hazardous waste disposal and many other things that have an impact on research institutions. Another reason is our recognition of the continuing need to harmonize the requirements of federal agencies. Harmonization is especially important in our application of existing animal welfare laws, regulations, and policies from the different oversight agencies involved. An additional reason is that obviously, it is important for us to "get it right" for the benefit of animal welfare as well as science.

INTERRELATED ISSUES

Although the real focus today is on pain and distress and the USDA Policy 11, which defines these, it is really impossible to separate Policy 11 and look at it independently from some other current initiatives because these issues are all interrelated. Each issue is very important in its own right but also because it has significant potential for unintended outcomes and burdens if not looked at along with others in an integrated fashion. I will comment briefly on some of these issues.

Similarity of Initiatives

The underlying principles for the USDA Animal Welfare Act and the Department of Health and Human Services Health Research Extension Act are remarkably similar, especially on the pain and distress issues. I discuss the specific elements of these documents below.

Principle IV. US Government Principle IV is a good example: "Proper use of animals including the avoidance or minimization of discomfort, distress, and pain when consistent with sound scientific practices, is imperative." I believe that statement is the simplest way of describing the "Three Rs" and the concept of alternatives.

The principle continues with that familiar anthropomorphic statement, which provides the starting point for making the necessary pain and distress assessments before determining how to avoid or minimize them. According to the statement, unless the contrary is established, investigators should consider that procedures that cause pain and distress in human beings may cause pain or distress in animals. It proceeds to address sedation, euthanasia, and other important points.

Another PHS policy (also NIH grants policy) that is not often mentioned is the requirement for investigators to address five very specific animal-related points in the vertebrate animal section of the NIH grant application. Study sections are expected to evaluate this information along with other components of the application. Applications without this information are considered incomplete.

In the statement at D, the principal investigator is asked to describe "procedures designed to assure that discomfort and injury to animals will be limited to that which is unavoidable in the conduct of scientifically valuable research...." These are just two of the dozens of references to pain and distress in the PHS policy and in the *Guide for the Care and Use Of Laboratory Animals* (NRC 1996). I mention them because they are not very different from the intended outcome of the USDA requirement to consider alternatives, that is, the infamous Policy 12.

Policy 12. Although far more prescriptive, the USDA language in Policy 12 calls for essentially the same assurances that are required in an NIH grant application. Dr. DeHaven and I have had this discussion, and I believe the revised Policy 12 reflects a more appropriate focus on what will be done to minimize pain and distress and what steps will be taken or were taken to learn about the availability of various alternative systems.

On this subject, some of you may be aware that our office is cosponsoring the production of an alternative search engine, working with the Johns Hopkins Center for Alternatives to Animal Testing and a number of other project team members. This search engine is now at the beta test stage, and we are very excited about its prospects. It is designed and intended to provide useful tools for investigators and IACUCs to make this process of considering alternatives not only much easier but also considerably more effective. We are looking forward to that development.

Policy 11. Dr. DeHaven has provided an excellent description of the possible changes to Policy 11. Of course, the definitions are extremely important because they determine not only what is reported, and where and how it is reported, but they determine also when Policy 12 might apply and play a role in decisions.

I do want to emphasize, as did Dr. DeHaven, a fairly major problem with the current Column E definition in which animals are assigned to Column E based on whether anesthetics or analgesics were withheld because they would interfere with the aims of the study. A very literal, technical application of this definition does not include the more important issue of whether the animal actually experienced pain or distress. I believe this area is of common concern for most of the participants in this discussion.

Another issue is that the current definition does not take into account the availability of many methods to alleviate pain and distress that are not classified as analgesics or anesthetics. For example, if a disease state is the cause of a painful or distressful condition, the effective treatment of that condition could obviously alleviate it. Alternatively, if a behavioral approach were used to alleviate anxiety or distress through something like training, conditioning, social housing, or simply comforting by a trusted caretaker, these methods of pain and distress alleviation would not count because they are not anesthetics or analgesics. The same description applies to other classes of drugs such as anti-inflammatory agents or antihistamines; they do not count for the purpose of reporting. This issue should be addressed so that the full range of pain alleviation methods can be credited. Unfortunately, some of these problems are embedded in the language of the regulations and may require a rule making to correct. I believe this area requires attention along with the changing definition and reporting requirements.

RATS, MICE, AND BIRDS

Another favorite topic is that of rats, mice, and birds. Dr. DeHaven has already alluded to this topic. You probably ask what this has to do with pain and distress. The PHS policy already covers all vertebrates, so there should be little change in the day-to-day care of these animals as a result of USDA coverage, at least at PHS-supported institutions. Our main public comment to USDA on this issue is that we seek consistency with the existing PHS standards and the *Guide for the Care and Use of Laboratory Animals* (NRC 1996). However, if one factors in modified reporting requirements, Policy 12 requirements, and possible definition changes for pain and distress, the sheer numbers involved could increase the administrative load on institutions very significantly.

HSUS PAIN AND DISTRESS INITIATIVE

I would like to make just a few comments about the Pain and Distress Initiative of the Humane Society of the United States. I doubt whether anyone in this audience favors increased pain and distress. Most of us are quite comfortable with the hope that some day it may be possible to achieve all scientific aims without pain and distress or, better yet, that we will have solved all of the medical problems that have relied on the use of animals so their use will become unnecessary. I think there is at least philosophical agreement on that point.

One of the concepts mentioned in the HSUS proposal, which I believe deserves discussion, is the idea of minimal risk. One of the HSUS proposals to USDA for altering the pain and distress categories is to establish certain thresholds below which particular levels of IACUC review may or may not apply. I have taken the liberty of extending that proposal perhaps beyond the point they had intended.

We do have a degree of flexibility now in the application of the designated versus the full review process. The parallel on the human subjects side for review, institutional review board (IRB) review, is a category called minimal risk. If carefully applied, this approach could free the IACUC's time to focus on much higher risk issues. However, neither the PHS policy nor the USDA regulations have provisions for minimal risk. Again, rulemaking would be required to implement such a suggestion. I believe the idea has some potential.

The current PHS policy and the USDA regulation, in contrast to the HSUS proposals, already require the minimization of pain and distress. The entire biomedical research community has already accepted this concept through the adoption of ethical principles by scientific professional societies, through journal editorial policies, and through commitments made as a condition for eligibility to receive PHS support. The concept of eliminating pain and distress is not very new. From the PHS perspective and expectation, it does not have to wait for 20 years.

CONCLUSION

In closing, I think the questions listed below cover some of the important public policy issues that I hope will be discussed further today and considered in the ongoing process of change in the regulations:

- Will proposed changes benefit animals?
- Are they consistent with statutes/regulations?
- Do they "harmonize" agency standards?
- Do they achieve goals while minimizing burdens?

REFERENCE

NRC [National Research Council]. 1996. Guide for the Care and Use of Laboratory Animals. 7th ed. Washington, D.C. National Academy Press.

Assessing Pain and Distress:
A Veterinary Behaviorist's Perspective

Kathryn Bayne

Associate Director
American Association for the Accreditation of Laboratory Animal Care
International

"Fundamental to the relief of pain in animals is the ability to recognize its clinical signs in specific species" (NRC 1996).

I thought I would set the tone for my presentation this morning with this quotation from the *Guide* because although today's objective is to provide federal regulators and policy makers with input regarding definitions of pain and distress, our larger goal as a research community is the prevention or relief of unnecessary pain and distress in laboratory animals. Although I will be discussing both pain and distress briefly this morning, the speaker who follows me, Dr. Gerry Gebhart, is much more knowledgeable about pain than I, so at the end I will offer up only a definition of distress.

ASSESSING PAIN AND DISTRESS

Carstens and Moberg (2000) suggest that a reasonable, although imperfect, approach to measuring the stress associated with pain and distress is to evaluate responses in four basic systems: the autonomic nervous system, the neuro-endocrine system, the immune system, and behavior. All of these systems reflect arousal of the animal; however, this morning I will focus on that area I know best—animal behavior.

The Dilemma of Definitions

As has been stated many times, much of the difficulty in achieving a broadly accepted approach to categorizing, and then addressing, pain and distress is due to the absence of a concise definition. From a behavioral perspective, this inability to arrive at a "Webster's Dictionary" type of definition is due in part to the fact that: 1) pain and distress are not discrete states, but are a continuum of experience; 2) signs differ between species, and most animals hide signs of pain because such a sign of weakness may provoke an attack from predators or subordinate members of the group; 3) there is a lack of specific behavioral indicators of pain; 4) interobserver variability can be large; and 5) there is a tendency to anthropomorphize, which is encouraged by US Government Principle IV. That principle states that "Unless the contrary is established, investigators should consider that procedures that cause pain or distress in human beings may cause pain or distress in other animals."

Issues with Assessing Pain

It is a well-established and accepted practice to use human experience to judge an animal's experience of pain and distress. Although I completely support this principle, I think we need to use this kind of assessment as a starting, not an end, point. We need to encourage investigators to include studies of pain and distress as they perform their research and to publish their findings.

Currently, there are several variables associated with pain assessment:

1. Assessments vary with the scale used, and they can be very subjective. What one person may view as a procedure that evokes moderate pain or distress, another may view as one that elicits minor pain or distress.

2. As Flecknell (1994) has noted, the absence of preprocedural scoring results in a lack of validation scores. There are no control data, so frequently confounding variables (such as those produced by analgesics) cannot be identified. For example, some of the consequences of surgery in rats, such as loss of body weight and suppression of food and water intake (signs frequently interpreted to be indicative of pain and/or distress), can also be produced in normal, unoperated rats by administration of opioid analgesics.

3. Chronic signs can be subtle and hard to detect. Changes in behavior due to pain and/or distress can be slow, incremental, and, individually, virtually undetectable. Whereas,

4. The dramatic, sudden onset of signs of pain is readily recognizable.

Issues with Assessing Distress

Distress is a generic term that can encompass anxiety, fear, boredom, frustration, and so forth. Thus, there are potentially multiple expressions of distress. Causes of distress that have been proposed include heat, light, sound, thirst, hunger, pain, novelty, exercise, pursuit, disease, and so there are also potentially multiple causes. But remember, the difficulties we continue to have in defining psychological well-being are due to a lack of direct reporting of the state of well-being by the subject (no language), not knowing what minor changes in well-being really mean to the animal, and not knowing what the impact is of levels of well-being on the research (which probably varies with the study).

There are advantages to using animal behavior as an assessment tool. For example, behavioral assessment is probably the least intrusive measure; and in the hands of a skilled, knowledgeable observer, behavior as an indicator of well-being (or lack thereof) can be reliable and powerful. A knowledgeable observer has the expertise to use different criteria in different species and to gauge their significance, and the skilled observer will predicate his/her assessment on the animal's milieu, including its physical environment, the research it is used in, and the animal's own status (e.g., its age and health).

However, there are also some disadvantages to using animal behavior as the assessment tool. Early experience (which may not be known), age, and physiological state can influence the behavioral response of an animal leading to inter- and intra-animal variability. For example, some research indicates that young animals are more responsive to pain stimuli than older animals and that sick animals may also be more responsive than healthy counterparts.

As has often been noted, correlations of specific pain-related behavior with intensity of the pain experience cannot be made. There is no behavior expressed by animals that indicates the severity of pain being experienced. Even the assessment of the efficacy of an analgesic is based on rather crude analgesiometric tests such as a hot plate or tail flick test.

There are too many amateur behaviorists, which can lead to overconfidence in their abilities. One lecture or even one course in animal behavior does not turn someone into an expert, just as the college courses I took in accounting and economics did not turn me into an Alan Greenspan. Our exposure to animal behavior through the popular press and television lulls us into a belief that we all have knowledge in the field and skills in understanding animal behavior.

Indicators of Pain

When evaluation criteria for pain are sought, a common approach is the use of general behavior that is extrapolated into indicators of pain in several species of laboratory animals. The problem is that this approach is rife with subjective criteria and/or contradictions. For example, Morton and Griffiths' (1985) seminal article, "Guidelines on the Recognition of Pain, Distress and Discomfort in Experimental Animals and an Hypothesis for Assessment," which has influenced numerous working group reports and reviews over the years, includes a table of specific, summarized behavioral signs indicative of pain and distress. Categories of behavioral criteria used by the authors are posture, vocalizations, temperament, locomotion, and other. Several of the behavioral criteria, such as dormouse posture, anxious glances, hangdog look, urgent squealing, and distinctive cry, are clearly open to a wide range of possible interpretations.

As recently as this year, a publication in *ILAR Journal* by Carstens and Moberg (2000) attempts to refine the Morton and Griffiths table by blending in elements from the NRC report *Recognition and Alleviation of Pain and Distress in Laboratory Animals* (1992); the FELASA working group report on "The Assessment and Control of the Severity of Scientific Procedures on Laboratory Animals" (Wallace and others 1990); and a *Laboratory Animal Science* article by Soma (1987) on assessing pain and distress. Carstens and Moberg propose different categories of criteria: general behavior, appearance, and physiology. This broader evaluation scheme may be more reliable than basing a judgment solely on one type of measure; however, it too contains some inherent difficulties for the evaluator. Specifically, the behavioral signs the authors report as indicators of pain in animals of the same species can be diametrically opposed; yet, both can be observed during experiences of pain. For example, the rat may express reduced activity or increased aggression; the guinea pig may squeal and stampede or go quiet; the dog may whimper, howl, growl or become quiet; and the cat may hiss or spit or become quiet. These inherent contradictions underscore the point that no single behavior is a reliable indicator of pain and that gauging pain requires a skilled observer not only to accurately record the signs but also to interpret them correctly.

It should be noted that behaviors used to assess pain are also not expressed exclusively during painful episodes (e.g., vocalizations). Stafleu and others (1992) make this point using the example of pig vocalizations. If you have ever worked with swine, you are aware that a minimum of restraint, sometimes even simply picking up a piglet, can result in very intensive screaming by the animal. Is the pig in pain? No. Stafleu and colleagues suggest that because piglets run a great risk of being crushed by their mother, they have evolved a low threshold for screaming to alarm the mother. The screaming might indicate some stress ("anxiety"), but even that behavior cannot be judged on this one measure. In addition, of course, the fact that pigs scream rather readily should not lead to the assumption that they are never in pain when they scream. Vocalizations are simply not a sufficiently sensitive measure to use solely in an assessment of pain.

Indicators of Distress

Indicators of distress are no less complex. Expressed behaviors that vary from the norm are generally used as indicators of distress. However, the observer must first establish what is normal for the animal (e.g., free-ranging vs. captive, aged animal vs. young animal) and then determine whether the behavior is not only atypical, but also maladaptive. This distinction is important because it is the expression of maladaptive behaviors that is likely to reflect a detriment in the well-being of the animal (e.g., high levels of locomotion vs. locomotion to the point where the animal is nonresponsive to potentially important external signals). Several scientific studies suggest that the expression of some atypical behaviors may be a coping mechanism of the animal either to increase or decrease its sensory input. Although this expression may signal that re-evaluation of the animal's environment is necessary, such coping activity may actually be an indicator of a lack of distress (Novak and Suomi 1988).

If the behavior profile expressed by the animal is narrow, it is important to identify which behaviors are lacking and whether their lack is a problem for the animal. One should ask what it means to the animal if certain behaviors are not performed. The animal should also be evaluated for changes in its behavior. Several small, incremental changes over time can result in an animal expressing significantly different behavior. Again, what does the change mean to the animal? Change in behavior should not necessarily be equated with behavioral pathology.

Degrees of Stress

As previously mentioned, stress is a continuum of experience. I would consider a mild stressor one that results in a short-term physiological response on the part of some animals and slight to no behavioral adjustment. Examples might include room entry and regular husbandry activities which have been shown to cause an increase in heart rate (Line and others 1989). A moderate stressor may include a minor procedure on the animal or a more significant procedure that is accompanied by pain relief and perhaps unconsciousness during the procedure. A moderate stressor would evoke behavioral adjustment on the part of the animal and physiological recovery or adaptation by the animal. The animal may experience limited distress (or perhaps eustress) associated with restraint. A severe stressor is one for which no relief is provided to the animal either through the ability to physically remove itself from the stressor or by modifications in its environment that would reduce the stress (e.g., use of nesting material to modulate cage temperature or treatment of a disease state). In such a case, there is inadequate adaptation by the animal to the stressor and distress results. I would, therefore, define distress as a state in which the animal is unable to adapt to the stressor and the animal may exhibit maladaptive behavior. The animal is not coping—behaviorally or physiologically.

Using Behavior to Assess Pain and Distress

So, how can we successfully use behavior as an indicator of pain and/or distress? I concur with the literature that recognizes two critical factors: (1) Individuals making the behavioral assessments must be knowledgeable and skilled in the interpretation of behavior; and (2) assessments should not be influenced by the personal biases of the observer (i.e., what the animal is perceiving vs. what the observer is feeling when observing the animal [Sanford and others 1986]).

Thus, behavior should be just one assessment tool used in the process by which pain and distress are assessed, monitored, and relieved. An overarching precept is that performance standards should be used in the assessment balanced with our scientific knowledge of animals' behavior.

A workable process for pain and distress management involves use of the IACUC. Specifically,

• The IACUC should review protocols for the appropriate use of pain-relieving agents.
• The committee should consider the criteria and determine a process for timely intervention, removal of animals from the study, or euthanasia if painful or distressful outcomes are anticipated.
• The committee's deliberations should be documented.
• The IACUC should ensure that a system to monitor and provide feedback is in place so that modifications in procedures can be requested as necessary.
• Guidelines should be in place that provide assistance in categorizing pain and/or distress for use by investigators and IACUC members.
• A classification system should be reasonable and consistently applied throughout the institution.
• A mechanism for prompt reporting of compromised animals should be developed and implemented.
• Personnel (research staff, animal care staff, IACUC members) should be trained in pain and distress management.

A Program of Pain and Distress Management

I believe our focus and that of the USDA needs to be on the experience of the animal. We should consider an entire *program* of pain and distress management, of which the animal's behavior will likely play a role in each of the following program elements: recognition, assessment, relief (if possible), and a feedback loop to animal care and use.

Responsibility for and oversight of animal well-being—and more specifically the minimization of pain and distress—is shared by several key institutional components: the research staff, the IACUC, the veterinarian, and other animal care staff. Their roles are separate, yet they overlap. There should be a flow of information and synergy among these components that results in a *strong* program to identify potential pain or distress-inducing circumstances, implement change (where possible) to minimize/eliminate the pain or distress, and include follow-up and monitoring to ensure that the goals of minimization are achieved. It is critical to avoid a splintered program in which these various institutional elements operate in different directions, rather than building and supporting a uniform, cohesive, and proactive program of pain and distress management.

REFERENCES

Carstens E., and G.P. Moberg. 2000. Recognizing pain and distress in laboratory animals. ILAR J 41:62-71.

Flecknell P.A. 1994. Refinement of animal use— assessment and alleviation of pain and distress. Lab Anim 28:222-231.

Line W., K.W. Morgan, H. Markowitz, and S. Strong. 1989. Heart rate and activity of rhesus monkeys in response to routine events. Lab Primate News 28:9-12.

Morton D.B., and P.H.M. Griffiths. 1985. Guidelines on the recognition of pain, distress and discomfort in experimental animals and an hypothesis for assessment. Vet Rec116:431-436.

NRC [National Research Council.] 1992. Recognition and Alleviation of Pain and Distress in Laboratory Animals. Washington, D.C.: National Academy Press.

NRC [National Research Council.] 1996. Guide for the Care and Use of Laboratory Animals. 7th ed. Washington, D.C.: National Academy Press.

Novak M.A., and S.J. Suomi. 1988. Psychological well-being of primates in captivity. Am Psychol 43:765-773.

Sanford J., R. Ewbank, V. Molony, W.D. Tavenor, and O. Uvarov. 1986. Guidelines for the recognition and assessment of pain in animals. Vet Rec 118:334-338.

Soma L.R. 1987. Assessment of animal pain in experimental animals. Lab Anim Sci 37:71-74.

Stafleu F.R., E. Rivas, T. Rivas, J. Vorstenbosch, F.R. Heeger, and A.C. Beynen. 1992. The use of analogous reasoning for assessing discomfort in laboratory animals. Anim Welfare 1:77-84.

Wallace J., J. Sanford, M.W. Smith, and K.V. Spencer. 1990. The assessment and control of the severity of scientific procedures on laboratory animals. Lab Anim 24:97-130.

QUESTIONS AND ANSWERS

DR. COUTO (Marcelo Couto, Scripps Institute and AALAS): Will AAALAC encourage institutions to use certified behaviorists to evaluate their enrichment programs, or will they trust amateurs as well?

DR. BAYNE: Although I am not here to speak on behalf of AAALAC, I am trying to discourage institutions from relying on amateurs without giving them appropriate training. There are many behaviorists who are not laboratory animal specialists. AAALAC looks at an institution's processes and at whether the outcomes conform with *Guide* recommendations and the other referenced resources that we list on our Web site. The Council does, in fact, encounter instances about which they feel compelled to comment, either as a mandatory item or as a suggestion for improvement regarding an institution's program of pain and distress management. They frequently include their observations on pain and distress management when they comment on the IACUC's operations.

DR. TAYLOR (James Taylor, NIH): I would only clarify that in looking at an institution's process, we actually look at the results of that process. If they appear to be inadequate or absent, then we are going to communicate with the institution.

DR. GEBHART (Gerry Gebhart, University of Iowa): Dr. Bayne, I inferred from your comments that you believe the definition for distress proposed by the USDA might be inappropriate. In my opinion, that definition is more likely to represent stress, rather than distress, of an animal. Is that a fair interpretation?

DR. BAYNE: Yes, thank you for highlighting that distinction. The USDA should review the language to reflect that the animal's inability to adapt or cope is distress due to the anxiety, fear, change, and novelty Dr. DeHaven described. All of those states are considered by ethologists to be stressors (not distressors). Only when an animal is unable to cope will it slide into the category of distress.

DR. GLUCK (John Gluck, Kennedy Institute of Ethics): You were very critical of the Morton and Griffiths papers, but I do not understand the nature of your criticism. Certain postures or eye movements appear to be relatively structural definitions that do not require a great deal of functional interpretation. Please describe your criticisms more explicitly.

DR. BAYNE: I believe the Carstens and Moberg paper is actually better because, with 15 years of additional information, it relies on the physiology of the animal in addition to behavior. The Morton and Griffiths paper, however, relies strictly on behavior and on the use of terms that are inherently subjective for assessing an animal's state of pain or distress. I am not sure I know what a "dormouse posture" is. I would have to go back into *Walker's Mammals* to see what a dormouse looks like. I am not sure we would all agree on an "anxious look." I think many Bassett hounds look anxious, and they are not; it is simply the expression. The amount of subjectivity in the criteria they proposed and the tremendous influence of the paper on many other, particularly European, positions are disturbing to me.

I believe David Morton has evolved his position and has, in recent years, developed very elaborate, detailed score sheets for rodents. However, they are not extrapolatable to all laboratory animal species. They, in fact, include other physiological dimensions of the animal's state.

Frequently, there is a great deal of reliance in the literature on such criteria as reproduction. However, if you deliberately preclude reproduction, then it becomes a moot point. Again, I would question the use of sole reliance on behavioral criteria because they vary from species to species. As I attempted to imply, there is even intra-animal variability based on an animal's experience as it ages, so our criteria must shift accordingly to avoid over- or underinterpreting.

DR. GLUCK: I am not trying to defend David Morton, but I think the overall impact of that paper and some of his other publications successfully draw attention to later, more formalized procedures whereby the experimenter, researcher, or staff member is required to check an animal's conduct in one form or another.

DR. BAYNE: I do not argue that point at all. I simply do not want institutions to go back and use those tables, which is what the authors were proposing.

DR. NELSON (Randall Nelson, University of Tennessee, ILAR Council, and IACUC member): You made a very good point, which is that we should assess the animal's behavior based on what is normal for that individual animal. How then do we deal with the assessment of who has the expertise to make the appropriate distinction between normal and abnormal? I submit that sometimes the investigator, although biased, may have much more knowledge about the animal's normal behavior than the veterinarian or the IACUC member.

DR. BAYNE: I agree, and if I were an IACUC member, I would not want to rely on just one source person. For example, the expertise of rodent caretakers is different from that of primate behaviorists. An institution probably has many, many different resources which the IACUC should use. With regard to PIs, they have typically studied certain animals in graduate school, they tend to work with the same species, and they know their animal models very well indeed.

Transgenic animals, of course, are developing a variety of behavioral profiles that are different from what we consider those of the standard mice. If that development is normal for that transgenic animal, then it becomes your baseline. As in any good scientific study, you need to evaluate your baseline. You need to be certain that your baseline is not changing and that you do not apply the same one to every study.

A good IACUC is going to be very proactive in probing, asking questions, and becoming very knowledgeable in their institution's talent and the relevant published literature.

Scientific Issues of Pain and Distress

G. F. Gebhart

Professor and Head, Department of Pharmacology
University of Iowa
Iowa City, Iowa

I am pleased to have the opportunity to follow Dr. Bayne's presentation because she anticipated many of the concerns and things I want to discuss. My charge was to comment on whether the principles and definitions that were written and presented in the 1992 NRC report *Recognition and Alleviation of Pain and Distress in Laboratory Animals* are still applicable today and still have merit. I believe they do, and I will discuss issues of pain (including hyperalgesia) and distress. I also will raise the concept of preemptive strategies with respect to anticipation of postoperative pain.

PAIN VERSUS NOCICEPTION

Pain

The widely accepted definition of pain was developed by a taxonomy task force of the International Association for the Study of Pain: "Pain is an unpleasant sensory *and* emotional experience that is associated with actual or potential tissue damage or described in such terms." A key feature of this definition is that it goes on to say, "pain is always subjective." This aspect of the definition reflects on the issue Dr. Bayne raised when she commented about interpretation of animal behavior and appearance by an observer based on feelings of the observer. We naturally have the tendency, when we observe an animal, to use our own past experiences to interpret and comment on what we perceive or believe to be the animal's status relative to discomfort, pain, or distress. It is very difficult, if not impossible, for our past personal experiences to be meaningfully applied to an animal. Training and experience in studying and observing animal behavior are required to interpret what we observe in nonhuman animals.

Nociception

The important distinction between pain and nociception must also be made. Nociception is the term introduced almost 100 years ago by the great physiologist Sherrington (1906) to make clear the distinction between detection of a noxious event or a potentially harmful event and the psychological and other responses to it. Sherrington and others before him understood that pain was not a simple sensation, but rather was a complex experience, only a part of which was sensory in nature. Accordingly, it is most accurate to describe what we study as pain in nonhuman animals as nociception. However, although nonhuman animals cannot express in words the psychological and emotional consequences of a noxious stimulus or event, none of us in this audience would hesitate to apply the term "pain" to that circumstance. This fundamental distinction between pain and nociception emphasizes the importance of interpretation of animal behavior by an experienced individual to assess the presence and intensity of pain and distress.

Having said this, it is nevertheless imperative to acknowledge that unless it is established to the contrary, we should assume that those procedures that produce pain in us might also produce pain in animals. This is an entirely appropriate guideline, bearing in mind the caveats that I discussed and the comments Dr. Bayne made just before me.

Noxious Stimuli

I want to consider briefly the nociceptive apparatus (i.e., basic anatomy and physiology). It is important to do this because our concerns about pain in animals are not related to acute pain associated with procedures that are of short duration and/or simple analgesiometric tests, as Dr. Bayne mentioned, such as a tail flick test or a hot plate test. Rather, we are more concerned about the consequences of procedures (surgical procedures included) that may be associated with longer lasting pain that may cause distress.

In Figure 1, a piece of skin that is innervated by a variety of sensory receptors called nociceptors is illustrated. Nociceptors respond only to stimuli that damage tissue or have the potential to damage tissue, and there are several kinds of nociceptors identified in Figure 1. When a nociceptor is activated by a mechanical, thermal, or chemical stimulus, the stimulus energy is transduced by the nociceptor to an electrical event (action potential) and the information is conveyed along nerve axons to the spinal cord, a part of the central nervous system. Input from nociceptors is transferred in the spinal cord dorsal horn to spinal neuron cell bodies whose axons ascend to supraspinal (brain) sites. Nociceptive

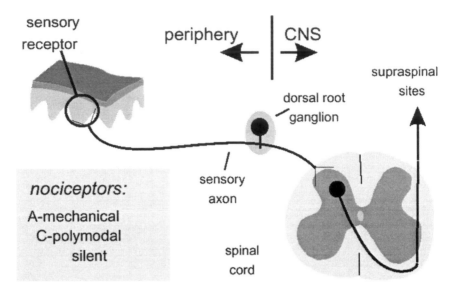

Figure 1 The receptors in the periphery that respond to noxious stimuli, here illustrated for a piece of skin, are termed nociceptors. There are three principal nociceptors: (1) Aδ mechanical nociceptors, which respond to noxious mechanical stimuli that damage or threaten to damage tissue; (2) C-polymodal nociceptors, which respond to noxious mechanical, noxious thermal (>44°C), and noxious chemical stimuli; and (3) silent (or sleeping) nociceptors, which do not respond to acute noxious stimulation of uninjured tissue but become active after tissue is injured. Information from nociceptors is conveyed by sensory axons, whose cell bodies are in the dorsal root ganglion, to the spinal cord where they synapse onto second-order spinal cord neurons, which transmit the information to supraspinal sites (e.g., the thalamus in the brain).

information is thus distributed to multiple brain sites that give rise to both simple and complex responses to the peripheral noxious event.

CLASSIFICATIONS OF PAIN

There are many ways to categorize pain. For example, pain can be classified in terms of duration. I think to classify pain in terms of its duration only is inappropriate; that is, to categorize it as acute or short-lasting as opposed to chronic and long-lasting is entirely arbitrary and not particularly helpful.

Protective Pain

I consider the diagram in Figure 2 to be a useful way to consider pain. The diagram suggests that we consider whether pain is "normal" in the sense that it evokes appropriate protective reflexes and/or behaviors. An example of an acute, protective, nociceptive reflex is withdrawing your finger from a heated surface you touch unexpectedly. If you analyze the event, you will note that you actually removed your finger from the heated surface before you perceived the event as painful. Another more common type of protective pain is that associated with tissue inflammation and tissue repair. One common example is postsurgical pain, which is longer lasting than the pain produced when one touches a heated surface.

Nonprotective Pain

There is also pain that I consider abnormal because it serves no protective value. Examples of these types of pains in humans are those associated with terminal cancer, nerve injury, and after a stroke. Some of the pain syndromes

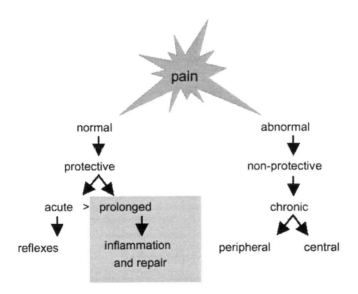

Figure 2 Pain that might be considered "normal" typically serves a protective function. Acute noxious stimulation, such as unexpectedly touching a hot surface or pricking a finger with a needle, evokes a nociceptive withdrawal reflex that, for example, prevents us from further burning our finger. After tissue injury associated with inflammation and repair processes, the injured tissue exhibits enhanced sensitivity to stimulation, termed hyperalgesia (see text and Figure 3). The enhanced sensitivity (e.g., to pressure applied to a sprained ankle or at the site of a surgical incision) is also normal and protective, preventing us from further damaging already injured tissue. Pain that might be considered "abnormal" provides no protective function. Nonprotective pains include those associated with cancers, peripheral and central nervous system damage (e.g., following a stroke), and others often classified as chronic. These pains certainly tell us that something is wrong, but they serve no protective function.

produced are bizarre and very long-lasting; some cannot be controlled and have no protective value to the organism whatsoever.

TISSUE INJURY

Sensitization

What are the consequences of tissue injury that might be associated with a surgical procedure? There is sensitization of the nociceptors illustrated in Figure 1. By sensitization is meant that the nociceptors change their behavior. They become more sensitive to stimuli that are applied to them. This factor contributes to what is termed hyperalgesia, which is discussed below.

There is also an awakening of so-called silent or sleeping nociceptors. Silent nociceptors exist in all of us and invest all of the tissues of our body. These silent nociceptors apparently play no normal physiological role in nociception (or pain), but when tissue is injured or damaged, silent nociceptors become active and begin to contribute information to the nervous system that was never previously contributed.

This activity is one way in which tissue injury or damage helps you protect yourself from further damage. As a consequence of these two events (i.e., nociceptor sensitization and awakening of silent nociceptors), exaggerated sensations can be provoked when you apply a stimulus to or near the site of injury.

Hyperalgesia

In Figure 3, the consequence of tissue injury and the operational definition of hyperalgesia are diagrammed. Plotted vertically on the Y-axis is pain sensation arbitrarily from 0 to 100; and along the horizontal axis is stimulus intensity, ranging from innocuous, nonpainful intensities into noxious or painful intensities. The normal range of pain-produced behaviors is described by the line labeled normal, which represents a normal psychophysical stimulus-response function. If I were to test everyone in the audience with a thermal stimulus or a mechanical stimulus of varying intensity, as the intensity of the stimulus increased, each of us would report increasing pain sensation associated with increasing intensity of stimulation. This stimulus-response function can be shown to exist for individual central nervous system neurons; individual afferent fibers innervating skin, muscle, joints, or viscera; and the behavior of nociceptors. This result applies throughout the realm of biology, if you will, from the single neuron to the integrated response that I would obtain from you when I ask, "Does this hurt?" and "How much does it hurt?" When tissue is injured or insulted in some way, the normal psychophysical function shifts to the left. The extent to which it shifts is dependent on the magnitude of the insult or the injury. I have illustrated in Figure 3 an example. In the normal individual, the noxious stimulus intensity

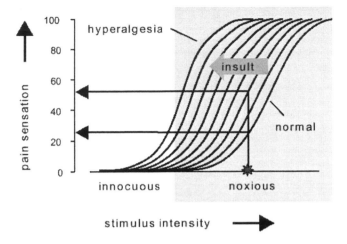

Figure 3 Illustration of how tissue insult affects responses to applied stimuli. Normally, there is a predictable response to stimulation that results in the sensation of pain only when the intensity of stimulation is in the noxious range. The line labeled "normal" represents a typical psychophysical stimulus-response function. In this illustration, a noxious stimulus produces a pain sensation of approximately 25 in uninjured tissue. After tissue insult, that same intensity of stimulation produces significantly greater pain (approximately 55 on the arbitrary vertical scale), which is referred to as hyperalgesia. Hyperalgesia is thus an enhanced sensitivity to noxious stimuli after tissue is injured. Note also that in some cases normally innocuous, nonpainful stimuli can produce pain after tissue injury, a component of hyperalgesia termed allodynia. Allodynia is most common after nerve injury.

identified by the symbol on the horizontal axis produces some amount of pain whereas in the presence of tissue insult or injury, the same intensity of noxious stimulation produces greater pain.

The two types of hyperalgesia are termed primary and secondary and are subserved by different mechanisms. Primary hyperalgesia is associated with a change in the behavior of nociceptors at the site of the injury. Secondary hyperalgesia is associated with a change in the behavior of neurons in the central nervous system. Obviously, the nervous system is not static; it is very dynamic and can be changed by tissue injury and insult. When we injure tissue in the periphery, a host of chemicals contribute to nociceptor sensitization and thus to primary hyperalgesia. These mediators (e.g., prostaglandins, amines, cytokines, kinins, peptides) provide targets for pharmacological intervention to prevent sensitization and thus to prevent the development of hyperalgesia.

When information from nociceptors arrives in the spinal cord, because it is contributed to by sensitized nociceptors and awakened silent nociceptors and is thus much more intense than normal, it increases the excitability of central neurons. This change in excitability is a consequence of increased release of a variety of other mediators and modulators (e.g., excitatory amino acids and peptides) released from nociceptor terminals in the spinal cord. These mediators contribute to an increase in the excitability of central neurons and to development of secondary hyperalgesia.

STRESS AND DISTRESS

Definition of Terms

The preceding description brings me to a discussion of some of the definitions in *Recognition and Alleviation of Pain and Distress in Laboratory Animals* (NRC 1992), which Dr. Bayne mentioned in the context of behavior. Stress is defined as the effect produced by external events or internal factors called stressors, which induce an alteration in an animal's biological equilibrium. I submit to you that getting up and speaking before an audience and not being able to get your computer to work is stressful. Stress is a normal component of life and does not produce distress unless it persists and leads to maladaptive behavior.

Distress is an aversive state in which an animal is unable to adapt completely to stressors and the resulting distress is manifested as a maladaptive behavior. The working definition of distress that Dr. DeHaven presented today, as I heard it, sounded more to me like a definition of stress than a definition of distress. One critical component of any definition of distress, because we cannot objectively measure or quantify distress, is behavior. As Dr. Bayne also indicated, we do not have reliable physiological measures that will tell us always that an animal is or is not stressed or distressed. We have no measures that provide reliable information to us about distress with the exception of the behaviors that we observe in these animals. Importantly, behavior must be assessed by individuals who are trained and knowledgeable about species' typical behavior and who understand what is normal and what is maladaptive.

Reduction with Preemptive Analgesia

One final point I want to make relates to preemptive analgesia. This strategy is based on the knowledge that there will be changes in the behavior of nociceptors and central neurons when tissue is injured. Thus, a preemptive strategy is particularly applicable to surgeries in animals. The strategy involves, for example, administration of opioids before the anesthesia is started and before an incision is made in an attempt to prevent the sensitization of nociceptors. The intent of this strategy is to prevent the development of central hyperexcitability and, as a consequence, reduce postsurgical hyperalgesia and reduce postsurgical stress and potential distress.

CONCLUSION

With that, I will close by saying that the definitions and principles presented in the 1992 NRC publication *Recognition and Alleviation of Pain and Distress in Laboratory Animals* are still valid and useful. I would comment that the working definition of distress proposed by Dr. DeHaven could be improved by more clearly including behavior, specifically maladaptive behavior, in the definition to clearly distinguish stress from distress.

REFERENCES

NRC [National Research Council]. 1992. Recognition and Alleviation of Pain and Distress in Laboratory Animals. Washington, D.C.: National Academy Press.
Sherrington C.S. 1906. The Integrative Action of the Nervous System. New Haven, Conn.: Yale University Press.

QUESTIONS AND ANSWERS

DR. KARAS (Alicia Karas, Tufts University): In my experience at a veterinary teaching hospital (Tufts University, Boston, MA), analgesics are commonly given to animals before surgery begins, and it does work. Interestingly, that observation was made several years ago at an American Society of Anesthesiology meeting, when the big-breaking media point was preemptive analgesia. One of my technicians expressed surprise when she heard about it on the Today Show because her experiences with preemptive analgesia included having animals recover quietly and with less distress.

DR. GEBHART: Thank you for sharing that experience. I suggest that many people who have tried this strategy and have reported that it docs not work have probably underdosed the animals. It is necessary to use a relatively large dose of drug to actually counteraffect the development of sensitization of nociceptors. When you do that, and I am glad your clinical impressions bear this out, you do see improved postsurgical behaviors in these animals. They appear to be less stressed as a consequence; they are up and about sooner, and they seem to recover much better.

DR. DE HAVEN (Ron DeHaven, USDA): In an effort to clarify the working definition that I used, distress is a state in which an animal cannot escape from or adapt to the external or internal stressors or conditions it experiences resulting in negative effects upon its well-being. In my discussion, I did go on to say that there are some things that may cause distress, and those things are boredom, anxiety, fear, pain. So although some stressors may lead to distress, that sequence was not intended to be part of the working definition. That list simply identified some stressors that can cause distress. My question is whether you were talking about some of those examples or were referring to the working definition that may inappropriately describe the distress and better describe stress.

DR. GEBHART: I was responding principally to the examples that were given as contributing to distress that, to me, contribute to stress. If they are prolonged, obviously, and the animal then cannot adapt to those stressors in the environment and begins to develop maladaptive behaviors, I would agree that every one of those stressors could lead to distress. I would suggest that "negative effects" be more clearly defined in terms of behavior, specifically maladaptive behavior.

DR. BAYNE (Kathryn Bayne, AAALAC): Dr. DeHaven, I believe you said that stress would have a negative impact on animal well-being. How is that negative impact going to be judged? Is the criterion going to be the expression of atypical behaviors or (what Dr. Gebhart and I are suggesting) that there is some evidence of maladaptation on the part of the animal?

DR. DE HAVEN: That is what we are here for, for you to answer that question for us. However, I seriously think there needs to be some evidence of negative effects upon the animal's well-being, and how we assess whether there have been those negative effects becomes critical.

DR. BAYNE: Yes, but I believe some people are going to assume atypical equals lack of well-being, instead of maladaptation reflecting a lack of well-being. Although you are not using either of those terms, the allusion to "negative impact on well-being" requires use of a method to assess the negative impact.

DR. HAYWOOD (J.R. Haywood, University of Texas Health Science Center): I believe we will have to involve stress physiology people to really start working on the development of measurements. However, we will have to address that at a later time. It is very interesting that when we start taking pharmacological approaches, we are going to be affecting all parts of the system and the response system.

The Humane Society of the United States Pain and Distress Initiative

Andrew N. Rowan

Senior Vice President for Research, Education, and International Issues
Humane Society of the United States
Washington, D.C.

Our pain and distress initiative consists of four main segments (HSUS 2000). First, we have commissioned the preparation of an expert report. Some may be beginning to wonder if it will ever arrive, and we are as anxious to see it as the rest of you are. Dr. Joy Mench, the chair of the panel, assures me that it is in process and close to completion. The report is meant to be a detailed analysis, somewhat akin to the NRC (1992) publication. The charge to the expert panel was given by providing an outline of the material we wanted addressed. However, the actual content will be the exclusive work of the panel.

Second, we have sent out five letters over the past 2 years to IACUCs as part of our outreach to research institutions. We are focusing on the IACUC as the institutional entity because we believe it has the appropriate oversight responsibility for this topic. We plan to send out more letters in the future. Some IACUCs have threatened to return our letters attached to bricks and less desirable objects that the US Postal Service might prefer not to handle, but by and large the reaction that we have had to date has been positive.

I know when I was chair of an IACUC at Tufts, getting the type of material that we have been distributing would have been useful to me and the other IACUC members. I am hoping that slowly people will begin to look past the source and find it useful to their IACUCs as well.

Third, we have also focused on regulatory issues and on some of what we perceive to be the shortcomings of the current regulatory system. Finally, we want to encourage more research and funding on how to assess and limit animal pain and distress.

CONTEXT OF THE INITIATIVE

Several of you in the audience have suggested that the message implicit in our initiative is too critical, especially of the laboratory animal personnel who are on the front line of laboratory animal care and welfare. We cannot escape the fact that the initiative does criticize what is currently being done, and I am not going to try to sugarcoat the issue.

However, I want to state very strongly that such criticism is not directed at people's motives or character. We recognize that there is an enormous amount of concern for animals in laboratories. Pain and distress is something that concerns all of us whether it is experienced by humans or animals. A concern for other beings' suffering is a fundamental element of any humane, empathic, individual.

What we think is the main issue is that we disagree on definitions and on the interpretation of what few empirical data are currently available to guide us in deciding what causes animal distress. That is where I see the disagreement, and if there are friction and unhappiness with the HSUS pain and distress initiative, we would like to see the debate focus on how we differ as regards definitions and the like.

DEFINITIONS OF TERMS

People commonly use pain, distress, and suffering as synonyms. I want to argue very strongly that these words do not refer to the same states. There are overlaps but they are absolutely not synonyms. There are other terms as well that describe parts of responses involved in suffering.

Twenty years ago, nociception was described to me as a scientific euphemism for pain. However, I now know that this is not a scientific euphemism at all. It is a very precise term that is not equivalent to the sensation of pain at all. It simply refers to the "stimuli" passing from the nociceptors up through the nerve fibers that conduct those stimuli. The sensation of pain may or may not be an outcome of such nociception. In fact, what may be viewed as behavior signaling pain may simply be a nociceptive reflex.

One example that illustrates this concept is the experience of a human being with a high spinal cord break. Such an individual could still have the nociceptive reflex loop for the lower limbs and will withdraw his or her foot from a hot iron or a painful stimulus, but there will be no pain perception because the central nervous system is not involved. The behavior is the result of a nociceptive reflex.

With regard to pain, the HSUS is not greatly concerned about pain research in the context of our pain and distress initiative. By and large, the International Association for the Study of Pain (IASP) has established good guidelines for ensuring that pain research protocols minimize pain. In quite a bit of acute pain research, the animal controls the level of intensity of the stimulus, and the animal says, "This is too much; I am going to stop now." In a sense, the animal is a volunteer as part of that program. There are concerns about chronic pain studies because the animal is no longer a "volunteer." In addition, pain caused in other research programs can be addressed using some of the newer long-acting analgesics.

Anxiety and fear are not the same as pain and, unlike pain, there is no obvious organic location associated with anxiety. These states can be treated with anxiolytics and sedatives, but I doubt that many programs do so. One example of the impact of fear in the laboratory was a dog model of anxiety developed by W. G. Reese (1979) out in the Midwest 20 years ago using pointers. In the presence of humans, the pointers would urinate, defecate, and display tonic immobility. If you turned your back on them, you would get bitten. However, in the absence of humans or under anxiolytic therapy, you could not tell which pointers came from the anxious strain and which ones did not.

Discomfort is different from pain and anxiety/fear. It may have an organic location, but we really do not have any sort of good treatment other than palliative care. Disease and malaise may involve discomfort but also include states that are different from those we might characterize typically as discomfort. Distress occurs when any of the states listed above (pain, anxiety, fear, disease) become too great. Distress may be prevented by appropriate psychosocial initiatives. For example, an animal that trusts its handler is far less likely to be distressed than an animal that is fearful.

Finally, we have the overarching concept of suffering. Suffering is a state that requires a certain level of pain, anxiety, or distress before it is experienced. As I discuss later, suffering and distress are not the same concepts. The model that may be used to describe the relation between some of the baseline states like pain, fear, discomfort and disease, and distress and suffering is as follows. The primary baseline states occur at different intensity levels and for different time deviations. Whether or not these baseline states cause distress is a function of intensity and duration. A low stimulus for a long time may cause distress whereas an intense stimulus for a moment may not. Once one has a state of distress, the level of suffering experienced is a function of the cognitive capacity of the being experiencing the distress.

Let us examine the concept of suffering in a little more detail. Some interesting anecdotal data in the literature describe a few humans (not many, it is a very rare syndrome) who cannot feel pain. When you read some of these case studies, you find that some of these individuals report that they are very fearful of surgery or that they just cannot stand going to the hospital. They fear the hospital and report that they suffer in hospitals, and yet they experience no pain whatsoever. This is one of those examples which proves the point that pain and suffering are not the same.

Another side of the suffering question (related to cognition) is the question of whether insects can suffer. A fairly persuasive argument can be made that they do not. Insects have short life spans and there is not a compelling evolutionary reason why pain-induced learning would make much survival sense from an evolutionary context for an insect.   Also, they do not have the central nervous system machinery that appears to be necessary to develop a sense of suffering (lobotomized humans have very low affect and do not appear to suffer as we know it).   By and large, one can explain most insect aversive behavior in terms of nociceptive reflex loops, rather than requiring a sense of distress or suffering.

ASSESSING DISTRESS

In summary, distress may be caused by pain; however, in the laboratory, it is probably much more often caused by illness, fear, or anxiety.   When does stress become distress, and what specific measures could help us determine when stress becomes distress?   This is an area where we are still completely in the dark in our opinion.   There are various types of empirical measurements that one can make at a distance, such as behavior or appearance. Other measurements can be made on handling, such as physical condition, weight, pulse rate, temperature, clinical signs, provoked behavior, and biochemical measures such as the blood level of the corticosteroids.

A whole raft of different physical and biochemical parameters can be measured that produce differences from one situation to the next.   However, none of these measurements let us know where to draw the cut-off points between mild, moderate, and severe distress.   Mouse blood corticosteroid levels have been measured at between 4 and 500/ng/mL.   Where between 4 and 500 do we draw a line and characterize the animal as experiencing distress?

Currently, the common approach is to say, "I looked at the animals, and they looked fine."   However, when we look at the animals, we usually look at them during the day.   Rodents are nocturnal animals.   The level of activity during the day is about 5% of the activity at night.   Therefore, if you are looking for differences, you are looking at a very low level of activity in the first place.   They are all asleep in the back of the cage, perhaps.   It is not easy under these circumstances to see whether they really are fine or whether they have some mild or moderate disturbance.   At night, or under red light (as is used at the University of Birmingham in England [D. Morton, 1997, personal communication]), it will be much easier to tell whether the animal is not behaving normally, but I doubt that many programs inspect the animals at night or under red light.

Norm Peterson (personal communication, 1999), now at Johns Hopkins University, acquired four cages with activity wheels attached and looked at nocturnal activity of the animals based very simply on the number of times the activity wheel turned every night.   In a normal cage of four mice, the activity wheel turned 1600 revolutions per night.   When the mice were given intraperitoneal injections of incomplete Freund's adjuvant before the end of the day, the activity wheel turned 900 times per night.   What does this difference mean in terms of distress?   Do we have any reason to be concerned by such a difference?   We at the HSUS would argue that we do, but perhaps others would argue differently.

Nonetheless, the argument is currently moot because we simply do not have much behavioral data on which to even start a debate.

Weight loss is a measure that we increasingly believe has a lot of promise as a screen for stress and distress. It is a relatively simple measure, and it is easy to obtain. Most stresses induce a transient decrease in feeding, an increase in the metabolic rate, and a negative energy balance due apparently to a biomechanism involving the actions of corticotrophin releasing factor and local cytokines (Dallman 2000). We have few data on weight loss in different experimental situations, and each strain of rat or mouse is likely to produce a different standard growth curve. In the HSUS white paper, we state that weight loss is a relatively insensitive measure. However, we retract that claim and now recognize that it can be very sensitive indeed. We will be pushing weight measurements as a simple way to promote both better science and less animal distress.

CONCLUSIONS

In terms of changes to the system, we absolutely must have adequate definitions. Dr. DeHaven provided a definition of distress this morning. Whether that definition is the right one is something we all need to discuss, as I am sure Dr. DeHaven agrees.

We must establish consistent reporting practices and report pain and distress when it occurs. The HSUS is concerned that if we are not reporting pain and distress when it occurs, then we are not paying attention to it. Of course, simply reporting such pain and distress does not guarantee that it will be attended to, but at least we would know that it was not being overlooked due to lack of attention.

We would like to see three grades of pain and distress—minor, moderate, and severe. The reason we would like to see reporting of all three grades is because we think we can focus our attention on severe distress and start studying how we can refine the technology to reduce that grade to moderate and then, eventually, to mild. Reporting severe pain and distress permits us to set priorities for research and action.

We need to support research on distress identification and alleviation. Such research might include more studies of weight loss and behavior in different research situations. The generation of more data then needs to be followed by broad-based discussions to determine cut-offs for the different categories. The fact that establishing such cut-offs will be difficult and involve some uncertainty is no reason why we should not start the process.

We need to support claims of action with real data. Such data is not something that the HSUS will be able to develop; such data will only be developed in laboratories and research facilities. Most of you already have access to such data anyway. We simply need to identify the personnel time to collect and analyze it.

We are pleased that attention to the pain and distress issue is growing. We believe that greater attention to these issues will benefit science, will benefit scientists, and of course will benefit the animals.

REFERENCES

Dallman M.F. 2000. Coping with challenge: Welfare in animals including humans. Dahlem conference paper. Berlin: Dahlem Foundation. In press.

HSUS [Humane Society of the United States]. 2000. HSUS Pain and Distress Initiative. URL: <http://www.hsus.org/programs/research/pain_distress.html>.

NRC [National Research Council]. 1992. Recognition and Alleviation of Pain and Distress in Laboratory Animals. Washington, D.C.: National Academy Press.

Reese W.G. 1979. A dog model for human psychopathology. Am J Psychiatry 136:1168-1172.

QUESTIONS AND ANSWERS

DR. GEBHART (Gerald Gebhart, University of Iowa): Dr. Rowan, regarding your discussion about developing categories of mild, moderate, and severe pain and distress, the issue is going to be who will provide the category. Is it going to be the investigator, the veterinary technician, or the veterinarian? I guess these are issues that have yet to be resolved, but I would like your opinion regarding who should be doing that and how you would define the different categories between mild, moderate, and severe.

DR. ROWAN: I would not even begin to develop the definition. This area is where we believe we need data and a detailed discussion between researchers, laboratory animal care staff, and animal protectionists to develop appropriate guidelines. Although IASP developed guidelines, it did not produce guidelines for categories of mild, moderate, and severe pain. Perhaps it is too ambitious a challenge to expect to produce reasonably clear categories. Nonetheless, some other countries are attempting to establish several categories of pain and distress, and they appear to be able to come to some level of agreement on the systems they use.

I would argue that leaving the development of appropriate categories up to individual institutions would be one way to start, but that there should be some sort of overarching data gathering group that develops some standard measures, once again based on data.

Part of the impetus behind the HSUS initiative is to stimulate a data-based dialogue and argument. I do not want to stand here and say 10% weight loss should classify a procedure under severe pain or distress because 10% weight loss occurs very quickly in a rodent. Perhaps it should be characterized as moderate, or even as mild, distress. Who knows?

We need to engage the issues and learn what people think and why they think it. What is needed is a lot more traditional scientific dialogue, which has underpinned so much of the biomedical progress of the past century.

Pain and Distress
Caused by Experimental Procedures—
Is It Time for a Reality Check?

Alicia Karas

Assistant Professor of Anesthesiology, Department of Clinical Sciences
Tufts University School of Veterinary Medicine
North Grafton, Mass.

As a veterinary anesthesiologist, I often speak about pain in clinical patients as well as laboratory animals. My experience is predominantly with pain caused by surgical interventions. Many years ago, at the time when we began to worry about having an IACUC, I was a laboratory animal technician. I am now an IACUC member and an investigator, so I am very aware of the problem of adding work or regulatory burden.

UP-FRONT REPORTING

In relation to whether the pain and distress classification systems are adequate, I think there is a different problem than just the wording. The problem I see is in the way people are complying with the policies and the regulations. I also believe that we are in some cases, as the HSUS contends, underreporting pain and distress and not adequately treating it because of the way we do things.

I think that most institutions probably suffer from an up-front reporting bias in actually recording the category, the number of animals, and the pain and distress category ahead of time. If you conducted a survey, you would find that most institutions probably use that information for their numbers at the end of the year, and it infrequently becomes altered.

Predictability

When the IACUC and the attending veterinarian decide ahead of time whether a given procedure should be in a pain category, the process suffers from limitations of predictability. I might know, for example, that a given type of surgical procedure causes a particular amount of pain, but because this is science, we are not always doing exactly the same thing. We might be drilling a hole in the femur this week and putting in something that causes more pain than in previous experiments.

In addition, as mentioned in the HSUS April 2000 report, there is no category for procedures that cause pain and distress that were partially but not fully alleviated with drugs. Of course, pain medicine specialists will tell you that you are very seldom going to actually get rid of every single iota of pain felt post-operatively by a subject having surgery. As someone mentioned earlier this morning, a few investigators might give aspirin to their subjects and consider it to qualify as "giving analgesics"; however, these cases do not address the concern of whether pain and distress are *adequately* alleviated.

Another problem of up-front reporting is that it blinds people to reality: because a study is assigned to Category D, it therefore will remain in Category D. None of my colleagues in the institutions where I have worked have thought we should go back and reclassify a study after a pilot run.

We should ask whose responsibility it is to validate the category assignment. As Dr. Rowan asked, is it a lot of pain? Is it a little pain? Is it the responsibility of the IACUC, the investigator, the attending veterinarian?

Planning for Optimal Animal Care

With regard to study design, anyone who has been involved in research knows that it is not possible to write a protocol that is 100% accurate. You can try your best to decide what you are going to do, how you are going to do it, how you are going to treat pain, and how you are going to assess pain; but you may have to change it. Pain can last longer than predicted. Events happen. Problems happen. Sometimes a procedure takes longer than is predicted. It is less possible than predicted. Because of those things, flexibility in the way the IACUC views the investigator's protocol is an absolute necessity for optimal animal care.

HSUS also states, and others today have mentioned, the following: "Animal pain, distress, and suffering are not easy to recognize or measure unambiguously. There is considerable opportunity for legitimate disagreement among scientists." I agree that pain, distress, and suffering are not easy to recognize. As stated in the *Guide* (NRC 1996), it is necessary to understand species-specific indicators of well-being. Pain and distress can be very difficult to predict and sometimes difficult to recognize; however, when we look at the animal, we are *often* going to find that it is necessary to revisit either the plan or the category assignment.

An example of this difference is the fact that two people can disagree. In my experience with a dog that had had a laparotomy to biopsy her liver, I worked with a gentleman who was a 20-year veterinary technician. He walked by this dog in the recovery room and said, "She looks great!" He was referring to the fact that she was sitting up, I think. She was not screaming or making any noise, and

TABLE 1 Type of Pain Expected from Surgical Procedures

Mild to Moderate	Moderate to Severe
• Ovariohysterectomy	• Thoracotomy / chest tube placement
• Extensive skin procedures	• Splenectomy / nephrectomy
• Laryngeal or oral surgery	• Significant resection of soft tissues
• Craniotomy	• Limb amputation
	• Orthopedic surgery
	• Eye surgery with corneal incision

he thought the dog was doing very well. I strongly disagreed and thought the dog was showing evidence of severe pain and/or distress. In fact, she was responsive to an additional dose of analgesic, and I saw a behavior change.

Based on my experience, I can tell you what types and what amount of pain are expected from certain surgical procedures (Table 1), but I sometimes have to rethink these expectations. Last week, I advised a student not to do an epidural on a dog to fix a lower limb fracture because I thought it would not be very painful. In fact, however, I was wrong. Again, flexibility is essential.

CATEGORY D

When we say that a study is in Category D and pain and distress will be alleviated by use of anesthetics, analgesics, or by other measures, do we then alleviate the condition? On our IACUC, we have been saying that if investigators do not give analgesics just because they do not know what to give and they do not know how to give it, they must put their study in Category E. Apparently, that course of action is not consistent with the policy.

I think that when animals still experience pain despite analgesic administration, it probably means that it is an ineffective choice of drug agent, or the wrong frequency or dose. It is necessary to change either the method or the category. I believe we need to have another category, that is, pain that was not able to be alleviated: Category D-2. This is simply an idea because I believe that most surgical interventions are not totally pain or distress free, even with medication. Although pain and distress may be largely alleviated, the biggest problem is that the investigator, and in many cases the laboratory animal veterinarian who might be the person who is consulted, is not likely to know how. I believe we need a lot of work in this area.

HSUS also states that "sensitive practical measures to gauge levels of distress in common laboratory animal species do not presently exist. For the most part, animal care staff rely on ad hoc observations or on relatively insensitive measures such as weight loss to ascertain whether animals are experiencing pain and/or distress." Dr. Bayne made an excellent (albeit heartbreaking) point that things are more complicated than we even suspected. However, I actually take the opposite view, which is that if we look at animals systematically, we *will* learn things. There is extensive literature on this subject, and evidence now exists for the behavior changes we see in animals. There is excellent work on how to assess pain in many different animal species, and I think it is possible to recognize much by careful study.

What then do we do? The method is not too difficult. Students scratch their heads and say, "I do not know how to tell if the animal is in pain." I would respond, "Did you look at him?" "Yes." "Did you touch him?" "No, I do not want to hurt him," they say. The fact is, it is simply application of scientific method, and we are, after all, scientists. One can read the available literature (the IACUC will know what is available), make regular observations, evaluate data, draw conclusions, and apply those inputs for the sake of the animals. Those steps are key to developing humane animal care methods.

To record observations, I have borrowed an idea from Dr. David Morton, and I know Dr. Hampshire has done the same thing. Using a laboratory dog assessment form (Figure 1), writing in a graphical or chart format is often better than scribbled notes on a sheet of lined paper. It is possible to see how things are changing over time and to get an idea of how long the animal actually needs to be medicated, have activity restricted, have appetite tempted, and so forth.

PLANNING EXPERIMENTAL METHODS

Planning of experimental methods should take into account that revisions of the methods may be necessary. If you have not used a surgical model before, try a first run-through in a surgical study in a pilot study. Record and use those observations in later studies. Include documentation of assessments and of any revision. If that information is available to the IACUC, and the IACUC wants to look at an investigator's study, they can request a formal monitoring sheet or assessment sheet such as Figure 1 to help them make their decisions even if they have not been in the laboratory seeing the animal. Documentation will help laboratory care personnel and other scientists and, of course, will benefit future work.

We found last week that we were working with some sheep that were about 100 pounds larger than sheep with which we had previously worked. Our plan simply to hold them, place a catheter, and give them an induction dose of anesthetic was not so easy because they outweighed all of us. We found that changing our method and giving them a premedication sedative, which is not necessarily commonly done, made things much easier on the sheep as well as on the scientists. Everyone benefits from something like that.

The IACUC must understand that assessing pain and distress caused by experimental handling procedures is a learn-as-you-go process in many cases. I have to disagree with Dr. Bayne's statement, that investigators are the experts on their animal models. In my experience, investigators are frequently coming from, for example, a contract research laboratory where they have been experimenting

Assessment schedule:

Date				
Time (am / pm)				
DISTANT OBSERVATIONS				
--- choose one --- 1. normal activity/sleeping/relaxed 2. inactive/lethargic/depressed 3. agitated/struggling/vocalizing				
--- choose one --- 1. normal breathing 2. increased breathing (effort/rate)				
HANDLING/EXAM OBSERVATIONS				
Rectal temperature / pulse (<100°F or >102.5 °F – contact vet)				
Capillary refill time / mucous membrane color				
--- choose one - **N - normal, A - abnormal** --- **N.** normal response to handling **A.** irritable / aggressive / hiding				
Able to stand / walk (**N** or **A**)				
Eating (**N** or **A**)				
Drinking (**N** or **A**)				
Coat soiled / wet (**N** or **A**)				
Sutures: swelling, discharge vs. clean/dry/intact (**CDI**)				
Urination (**N**ormal or **A**bnormal amount/color)				
Stool (**N**ormal or **A**bnormal amount/color)				
Other observations:				
Initials of observer				
PI or Vet contacted? (**Y** or **N**)				

Contact PI or DLAM veterinarian if you have logged: - **A**bnormal item in shaded row
- abnormal activity / breathing / handling behavior
- abnormal or no urination / stool / vomiting

FIGURE 1 Laboratory dog assessment form.

with rats, and they want to use a procedure in a different model. Alternatively, the investigator may actually know the animal from a biochemical and on-paper point of view but not from an observation point of view. In fact, in many situations, investigators may see animals only during procedures, and the animal care staff and the veterinarian are the people who are watching the animals. I therefore disagree with that premise.

Although I believe that planning ahead helps, it cannot provide for all contingencies, no matter how careful you are. There must be some feedback loop, an internal process for reviewing the pain and distress caused or alleviated and the measures taken. I believe the problem is not so much with the *definition* (although the regulators will claim to need it for their legal basis) but in the *implementation* and *outcome*.

The HSUS report also mentions that laboratory personnel can develop distancing mechanisms that help them cope but that often lead to people overlooking pain and distress. I want to make the point that distancing mechanisms are really a poor substitute for refinement. We could have not cared about how the sheep were doing, but in finding a better sedative for them, everyone was happier.

One major problem is the "disconnect" between the scientific community (including myself and others here from the scientific community), the investigators, and those who are developing and refining techniques. Are the investigators responsible for reading the refinement literature? They will say they have too much to do, and it is an admittedly difficult problem.

I think this disconnect is something that can be helped by training laboratory animal personnel in these methods, ensuring that they have a solid foundation. I actually took aside a college student who was working in our facility for the summer, and in 20 minutes, I was able to teach her to recognize more of pain and distress (pain in particular) than the investigator had led her to understand.

SCIENTIFIC JOURNAL REPORTING

HSUS also states correctly that there is limited published information about animals' experience of pain, distress, and suffering caused by typical laboratory procedures. I believe it was Dr. Zurlo, in a talk last year, who called my attention to a paper by Jane Smith and colleagues (Smith and others, 1997), which surveyed the biomedical literature and found that frequently you cannot tell what was done to animals by reading the published reports. I think that is a real problem because where else are the investigators getting their information?

In reading biomedical journals (which I now do fairly critically), I find many times that things were done and there is no mention of analgesics; there is no mention of whether the study was approved. I looked at one veterinary journal and found a paper written by at least two veterinarians in whose study dogs were anesthetized by mask isofluorane. They may have had very different dogs from those in my experience, because a dog with no premedication is quite difficult to mask down without a struggle. Then the authors had implants placed and to do that, they tunneled them from the neck to the flank and then turned them over, opened the abdomen, and implanted them into organs in the abdomen. Recovery was uneventful. That was it. No mention of analgesics or even whether the animals were euthanized at the end of the study. You can see that there is a big problem with what the authors actually reported doing to the animals. There is no mention of whether a few animals were lost due to problems with the technique in the beginning.

I would like to propose that when you review papers, you look for descriptions of animal care that include anesthesia and analgesia techniques. Look for specifics of acclimation and disposition and things such as whether animals were conditioned to be able to accept certain parts of the study. I believe that if we are not going to have these methods published, if we lack space that is precious in our research publications, then we must find some other place for that information. Many people are developing very interesting techniques to which we have been unable to have access. Our Internet user groups have made a huge difference because we can all speak to each other instead of talking on the telephone. We can speak to each other in large groups (and search previous conversations).

SUMMARY

The question in my mind is whether we should change the practice of upfront reporting of prospective pain and distress categories. I think the limitations of predictability can be overcome by appropriate observation and flexibility. Observation and documentation are key. If significant efforts are not made to assess and document pain and distress, then accurate reporting will never be possible. You will not know. No matter what method is used, we all might as well go home and not bother to redefine things anymore.

I believe that some sort of internal review must take place, and it can provide a reality check. The IACUC or the attending veterinarian, depending on the institution, can assume this role. Validation is achieved by periodic or even random (e.g., USDA inspections) IACUC review and consideration of category assignment. I realize that in proposing this, I am proposing that someone will be required to do more work.

I agree with HSUS that it is not beyond the scope and responsibility of the scientific community to determine underlying principles of pain and distress alleviation in animals, which can then be applied to various models and methods. I think it is not beyond our scope. It is going to take some work, but we can start right now by looking at the animals, at how they are doing, and documenting that information. We cannot say that everything went alright unless we check to see *whether it did.*

REFERENCES

NRC [National Research Council]. 1996. Guide for the Care and Use of Laboratory Animals. 7th ed. Washington, D.C.: National Academy Press.

Smith J.A., L. Birke, and D. Sadler. 1997. Reporting animal use in scientific papers. Lab Anim 31: 312-317.

A View from the Trenches

B. Taylor Bennett

University of Illinois, Chicago, Ill.

I have divided my remarks into four sections. First I would like to let you know who I am and what I believe qualifies me to be here making these remarks. Next I will explain how it is that I came to be standing here today. I will then explain why I cannot support efforts to introduce new language to the current definition of pain, the addition of a definition for distress, or changes in the current annual reporting process by revising the current Animal Welfare Policies. I will close by telling you why I think that any expenditure of fiscal or human resources to develop new paper definitions is not the best use of those resources.

WHO I AM

I have spent more than 30 years as a laboratory animal veterinarian in a large academic institution. The program of laboratory animal care that I have directed for the past 22 of those years is a program in change. It has always been a program in change. The changes have and continue to be directed toward the constant improvement of the program to assure that the animals in our charge receive daily care that meets or exceeds the recommendations of the *Guide for the Care and Use of Laboratory Animal* (NRC 1996), while at the same time providing the investigators in our institution with a working environment that supports their efforts to conduct high quality research.

I provide this background information to point out that even as a somewhat senior member of our specialty, I have not become so set in my ways that I routinely oppose change. I actually seek it and encourage change that will add to our knowledge base and improve the quality of care that laboratory animals receive.

HOW I GOT HERE

I am here as a representative of the American College of Laboratory Animal Medicine (ACLAM). The key phrase here is "as a representative." I am neither an elected officer nor a current member of the board, so I do not represent the leadership or membership of ACLAM. I am completing a 4-year term as one of the original members of the ACLAM Foundation. The Foundation's goal is to increase the body of knowledge in laboratory animal medicine and science by raising money to support research that accomplishes this goal. I also currently serve as ACLAM's nominee on the 1999 American Veterinary Medical Association's Panel on Euthanasia, and I previously served on the 1993 Panel. It is because of my current involvement on this panel that Dr. Margaret Landi, ACLAM President, called and asked if I could attend this meeting in her absence.

WHY I CANNOT SUPPORT CHANGES IN THE EXISTING POLICY

The USDA Animal Care Policies are in reality de facto regulations. They are interpretive rules that, although not legally binding, leave us only a challenge in court if we do not agree. There is no mandated public comment period with a subsequent response to those comments as part of publishing the Final Rule. As such, the regulated community, the scientific community, has no mandated opportunity to provide their input or to learn how their input was considered in developing the Final Rule. The policies in effect regulate a scientific activity, without having had the mandated participation of the regulated community in their development. Any changes to existing definitions, addition of new definitions, and revisions to the annual reporting process should take place as changes to the Regulations with publication in the *Federal Register*, a mandated public comment period, and publication of the Final Rule with USDA commenting on why it accepted or rejected those comments.

Why would I ask for more regulations? I am not asking for more regulations, I am asking for a process that gives the scientific community a full opportunity to address the available scientific information on pain and distress and its application in clinical practice. I believe that available scientific information strongly speaks for not adding additional language to the regulations/policies, and I believe that a formal public comment period will make this fact abundantly clear.

Pain and distress are responses to environmental and internal events that affect the well-being of those that experience them. Adequate definitions/ explanations can be found in the literature. Such definitions should be incorporated into the policy and procedure documents of every IACUC in the country. They do not need to be redefined in regulations and policies.

Pain and distress are medical problems that first require detection, then determination of a cause (the diagnosis), and then alleviation (the treatment). Pain and distress are medical care issues that must be dealt with on a case-by-case basis by a veterinary staff that must apply its didactic and experiential knowledge in managing them. These issues are fundamental to the success and credibility of any program of adequate veterinary care.

In my opinion, one of the most important sections of the Animal Welfare regulations is Section 2.33, Attending Veterinarian and Adequate Veterinary Care. This section in part states, "Each research facility shall assure that the attending veterinarian has the appropriate authority to ensure the provision of adequate veterinary care and to oversee the adequacy of the other aspects of the animal care and use." The institution must make available appropriate facilities, personnel, equipment and services to allow the veterinary staff to use appropriate methods to prevent, control, diagnose and treat diseases/medical conditions.

An integral part of this program is providing guidance to principal investigators about issues that have an impact on the well-being of animals being used in research. We all know individuals we turn to for guidance and direction. Why do we rely on these individuals for guidance? It is really quiet simple. We respect them because they have established their credibility with us, and we recognize their professional competency. The key to the success of a program of adequate veterinary care is the mutual respect of the veterinary staff and the investigators, which leads to a confidence level that makes providing guidance the underpinning of the program of adequate veterinary care.

The key to the success of a laboratory animal veterinarian in an academic environment is earning the confidence of research colleagues, and we do this by using scientifically sound approaches in dealing with issues involving the use of animals on a given protocol. Our colleagues are accustomed to searching for answers that can be supported and defended with scientific data. With this approach, the necessity of having to implement USDA regulations and policies that are inconsistent with available scientific information compromises our ability to do our job.

Because I am already on record concerning how the existing language in Policy 11 undermines the ability of laboratory animal veterinarians to do their job, by forcing us to implement policy that makes neither scientific nor common sense, let me give you an example of why I fear additional changes in the language. Included in the information that I was sent a couple of weeks before this meeting was a 1995 draft of a HSUS document entitled, "AWA Classification of Pain and Distress in Animal Research: A Proposal." That document includes a reference toVeterinary Service Memorandum 585.2, which provides guidance on pain and distress to area veterinarians in charge. In that memorandum, chronic pain is defined as follows: "Chronic pain results from a long-standing physical disorder or emotional distress that is usually slow in onset and has a long duration. It is seldom alleviated by analgesics but frequently responds to tranquilizers. . . ." The service memorandum proceeds to define anesthesia as "complete unconsciousness." It is these types of definitions that must be aired in a mandated public forum, before they establish the regulatory environment within which we laboratory animal veterinarians must operate.

In reading that draft document, it brought back memories of preparing written comments on the proposed regulations to implement the amendments to the Animal Welfare Act. At that time the USDA proposed language on painful procedures that they did not adopt after the public comment period. To try to adopt a similar system at this time without changing the regulations makes no sense.

It has been almost 15 years since this issue was first addressed. Why address it again, and why try to do so by changing a policy?

- Have things not changed in the last 15 years?
- Has the well-being of laboratory animals not improved in the past 15 years?
- Has the number of trained laboratory animal specialists who can provide adequate veterinary care not grown dramatically?
- Lastly, has not the investigative community not changed in the last 15 years?

Yes, Yes, Yes, and, most emphatically, YES!

Let us look at that emphatic YES. Today's laboratory animal veterinarians deal with a population of investigators who more and more are made up of scientists who were trained and built their scientific careers since the 1985 amendments.

- They accept the role of the IACUC.
- They accept the role of the veterinary staff in providing guidance.
- Most importantly, they accept the fact that the well-being of the research animals that they use is critical to the success of their research.

At a time of tremendous advances in biomedical research, unprecedented growth in the federal research budget, and tremendous competition for those dollars, the new generation of investigators does not just accept the need for quality animal and veterinary care programs, they demand it. They do not need to be reminded, by the development of regulatory definitions for pain and distress and increasing the record-keeping required to complete the annual report, that pain and distress should be minimized. It is a scientific necessity. Minimizing pain and distress is a way of life in designing and implementing research protocols, a process that involves the collaboration of the veterinary staff and the investigators in an environment that relies on the latest scientific knowledge and technology. Cutting-edge science requires cutting-edge support programs and facilities.

One of the things that makes laboratory animal medicine an exciting profession is the science that we help advance. So any definitions that affect how we function should be based on scientific data and be clinically relevant. Incorporating ever-changing scientific information into the regulatory process is no small task because the regulatory process is by design slow to get from point A to B, and scientific information is changing daily. The regulatory process is slow because of the many steps required to change regulations in our democratic society that protect the regulated community from counterproductive regulations. Developing new regulations to define pain and distress would be counterproductive at this time. It would slow the progress that our new generation of laboratory animal veterinarians and investigators are making in recognizing and managing pain and distress. Rather than spend our time trying to implement expanding regulations in this area, let us spend our time developing techniques and methodology that minimize pain and distress. Let us not waste our time discussing how to define pain and distress on paper. Let us spend our time defining what pain and distress is and how to manage and alleviate it.

Contrary to the good intentions of those who want to eliminate pain and distress from the lives of laboratory animals, laboratory animals must live in the real world, where there is no hope of eliminating pain and distress in our lives, although through research we can certainly find more and better ways to minimize it and its impact on our well-being. This knowledge will certainly lead to even more dramatic changes in the way we laboratory animal veterinarians can manage pain and distress. The progress made in laboratory animal medicine in the last 15 years has been significant. Let us all work together to continue that progress and not waste our time developing new regulations that not only will not accelerate this progress but may also serve to slow it down.

The thing that will increase this progress is more research, which brings me to my final point. But before I move on, let me remind you again that I do not routinely object to change. In fact, as I said, I usually encourage change. So if change is needed to address the issue of minimizing pain and distress, let us amend the Animal Welfare Act. Specifically I would propose an amendment to Section 13(a), (3), (C) to change its current wording, "in any practice which could cause pain to animals," to read, "in any practice which could cause pain and distress to animals," and add: "(i) that a doctor of veterinary medicine is consulted in the planning of such procedures."

This language would make it clear that pain and distress are medical conditions that should be dealt with by medical professionals. This language would clearly show the public that Congress recognizes the importance of providing adequate veterinary care and has mandated an expansion of the programs of adequate veterinary care to address the issue of distress.

WASTING OUR FISCAL AND HUMAN RESOURCES

On Sunday, June 11th, I had to carefully manage my time, because I had to organize my thoughts for this meeting and review the information I had received for the scheduled conference call of the ACLAM Foundation to discuss our reviews of 18 proposals that were seeking funding at the level of $15,000 each to address key questions in laboratory animal science and medicine. Of these 18 proposals, 11 clearly addressed questions related to defining and/or managing pain and distress in laboratory animals. In fact, 15 of the proposals addressed issues of refinement and reduction, and the other three addressed issues related to the safety of the work environment.

Just look at all of us here today. How many people had to travel to come to this meeting? How many spent at least $500 getting here? More than a $1,000? How many of you would ask for more than $10.00 an hour if you were applying for a new job? $20, $30, $40, $50? Obviously, there is enough money being spent now by our respective organizations to support an ACLAM Foundation grant. We could fund only seven this year. What if we could double that number, afford to give bigger amounts, be able to support multiyear grants? The number of questions that could be asked to find the answers to better define pain and distress for real, not just on paper, could be dramatically increased.

The ACLAM board has supported the foundation by committing more than 12% of its operating budget. The membership has provided and is committed to providing hundreds of thousands of their personal dollars to support research to improve animal well-being.

Instead of spending our time and resources developing and debating paper definitions, let us make a commitment to support the research that produces information to make it possible for the laboratory animal community to better recognize, manage, and alleviate pain and distress. To persist in this debate in lieu of really "putting our money where our mouths are" is, in my opinion, counterproductive and shows little real commitment to improving the well-being of animals that must be used in research.

So, in summary, I stand before you, somewhat reluctantly, as a member of ACLAM. I have been an active participant in laboratory animal medicine for more than 30 years. During that time, I have seen and hopefully participated in many changes that have improved the well-being of laboratory animals by improving the knowledge base of those who care for and use laboratory animals. Today we operate in an environment in which the majority of the care of laboratory animals takes place in programs managed by trained laboratory animal veterinarians working in concert with a new generation of investigators. Together we work on the common goal of reducing pain and distress in laboratory animals because it is the right thing to do and it improves the quality of science. In this environment, we do not need more regulations, we need more information; and I challenge the regulatory, animal welfare, and scientific communities to support research that provides the information that will ultimately lead to our ability to realistically address the issue of pain and distress.

REFERENCE

NRC [National Research Council]. 1996. Guide for the Care and Use of Laboratory Animals. 7th ed. Washington, D.C.: National Academy Press.

QUESTIONS AND ANSWERS

DR. DELL (Ralph Dell, Institute for Laboratory Animal Research): Let me begin with a question related to your statement that we are working hard to manage pain and distress in laboratory animals because we are good people and because it results in good science. There are critics who concede that most of you are good people but that a few are not. How are we going to identify the latter group unless we have some sort of an inspection system?

DR. BENNETT: I do not have a problem with an inspection system. I have a problem with trying to define basically through the regulatory process what is scientific and needs to be defined by the scientists and change as the science changes. That ability to accommodate change is very difficult to do with regulations. I do not think that one has to change regulations every time we find a negative situation. We simply find better ways in our institutions to deal with those instances.

I believe the vast majority of people with whom we work have one common goal, which is to minimize pain and distress. Even if we were not saying that we are good people, it is better science. Practicing scientists cannot afford to waste money creating unnecessary pain and distress in animals. They are under tremendous pressure to produce and they cannot afford to lose animals due to improper management.

DR. DE HAVEN (Ron DeHaven, USDA): I appreciate the fact that we get the full spectrum of positions at these public forums. I would therefore like to clarify two points. First, I heard you say that VS Memorandum 595 dates back to 1995; however, it actually goes back to 1985. That memorandum is from the prior organization.

DR. BENNETT: I certainly agree.

DR. DE HAVEN: The other, more important clarification relates to two initiatives that are under way, one of which involves a potential change to the policy and the other a potential change to the regulation. I think we have been remiss in the past in terms of all of the requirements for minimizing pain that did not include minimizing distress. I think that omission is clear in the act and in the regulation. Any update to Policy 11 will be primarily to include distress and to clarify what kinds of things constitute distress and therefore what kinds of activities would require consideration of alternatives to distressful procedures since the current policy and our practices in the past have focused on pain. As you point out, the change to the policy is the one interpretive rule.

Any change to the pain or pain and distress categorization would indeed require a rulemaking change. Any inclusion of a definition of distress in the regulation would require a rule change. It is either a two- or a three-step process for the rule change, which requires, at a minimum, publication of a proposed rule, public comment period, review of those comments, and publication of a final rule. In this case, we will start with a three-step process but include in the beginning an advanced notice of proposed rulemaking and request for comments. The next step, if we decide to proceed, would be a proposed rule. There will, indeed, be ample opportunity for the scientific community and all interested parties to have some input into that process.

DR. BENNETT: I believe it is important to note that unless a rule appears as a formal regulatory change, we do not get as much of the scientific community involved in the comment period because they expect these to be adopted ultimately. I formed that impression when I worked with organizations to write responses that appeared in the Federal Register.

DR. HAMPSHIRE (Victoria Hampshire, Advanced Veterinary Applications): Are you in fact supporting a revision to the regulation and/or the definition of an attending veterinarian?

DR. BENNETT: No. I am saying that the role of the veterinary staff would be strengthened in terms of the issue of distress if it were clearly pointed out that investigators should consult with a veterinarian when a proposal will have the potential for creating not only pain, but also distress. The law does not state that now. It only refers to pain.

DR. GEBHART (Gerald Gebhart, University of Iowa): I want to remind people that the way we have been talking about pain and distress seems to suggest that they go hand in hand. However, pain does not inevitably lead to distress; and pain, as Dr. Rowan said, is not a major concern in many respects because it may be addressed with the appropriate use of analgesics. By establishing a category that includes both pain and distress (Category 1, 2, and so forth), even if there were some pain but no distress present, or if only distress were present, an investigator's rating would imply that both conditions were indeed present. I believe we need to be careful about associating those two conditions as always being related because they are not necessarily so.

DR. BENNETT: One of the points made previously that I found interesting was that these reports are different from retrospective reports. I cannot imagine that any institution could complete that report without some type of retrospective reporting process that involves looking at the animals and evaluating these things.

DR. HAYWOOD (J. R. Haywood, San Antonio): We have discussed these reports as documents that actually provide us with information. However, maybe we should think about why we are actually filling out these reports. What function do these reports truly serve to help the animals?

MS. LISS (Cathy Liss, Animal Welfare Institute): I am delighted with what appears to be current support for the 85 amendments to the animal law and actually a cry for a bit of strengthening of those 85 amendments. I am delighted to hear that coming from you, Dr. Bennett, and that support does appear to be a change.

DR. BENNETT: My position has not changed, Ms. Liss. I have always supported reasoned and well-founded regulations.

MS. LISS: Finally, you talked about the expense of people coming here and being involved in this kind of process. All I can point to is at countless AALAS annual meetings where there are extensive receptions in which all researchers are eating, drinking, and having a good time. Perhaps those monies could be applied to enriching the lives of animals in laboratories.

I would also like to commend USDA for all of the input they are incorporating into changes in their policies. As you have pointed out, former veterinary memoranda and policies were very secret and difficult to obtain, and there was typically no outside input. There has been a significant change, and my hat is off to USDA for a very open process, for accepting this input, and for allowing time for that scientific input.

DR. GLUCK (John Gluck, the Kennedy Institute of Ethics): It seems to me that you were arguing earlier that economic factors in research, the competition inherent in good science, and the virtues of the investigators are adequate motivators to ensure proper concern or treatment of animals.

DR. BENNETT: I am describing the real world. I am arguing that instead of motivating them, those factors comprise part of reality, and that part of reality makes our job as laboratory animal veterinarians sometimes a lot easier to do.

DR. GLUCK: I still think you are saying that these factors influence good conduct on the part of investigators. I am saying that these things have been in place before now.

DR. BENNETT: I think the things that influence the conduct of the investigators with whom we work today are different from when I first got involved in this field. People then had been brought up in a different era. Sensitivities toward the issues of animal pain and distress today are totally different than they were 15 or 20 years ago.

DR. GLUCK: I am simply saying that these factors have always been in place. It has been the addition of public interest and regulations that I think has helped move that motivation.

DR. BENNETT: I am not denying that premise. One of the results of public regulation was to require institutions to provide us, the laboratory animal veterinarians, with the resources we need to do our job.

AALAS Position Paper on the "Recognition and Alleviation of Pain and Distress in Laboratory Animals"

Marcelo Couto

Scientific Advisory Committee
American Association for Laboratory Animal Science

Associate Director, Comparative Medicine Department of Animal Resources
Scripps Research Institute, La Jolla, Calif.

I want to thank ILAR and NIH for jointly organizing this workshop. This meeting is a great opportunity for scientists, regulators, and animal care representatives to come together and present their views on the important issue of recognizing and alleviating pain and distress in laboratory animals. I hope that at the end of the workshop, we will be better equipped to do a good job.

AALAS has produced this position paper to outline the difficulties in recognizing pain and distress in laboratory animals, to clarify certain definitions, and to stimulate the laboratory animal community and regulatory authorities to find appropriate means for addressing this problem and create lines of communication for sharing information that may advance this cause. We also urge the regulatory agencies to revise the current pain and distress categorization system in the USDA annual report.

EVALUATING PAIN AND DISTRESS

The assessment of pain and distress is a very individual matter, both from the standpoint of the individual animal and the observer. When evaluating pain and distress, it is very difficult to make general statements because different species and different individuals within a species often exhibit different thresholds and different tolerances to pain. Determining what constitutes pain or distress in animals is further complicated by the fact that there are no universally agreed-on criteria for assessing or determining what is or is not painful or distressful to an animal.

An additional, complicating issue is that pain and distress, as well as the drugs used to alleviate them, can introduce a variable in a study. Limited pilot studies and well-controlled experiments may be helpful in sorting out any confounding effects caused by either pain or by the pain-relieving drugs.

REPORTING ANIMAL USE

The complex nature of modern animal experimentation requires accurate reporting of animal use. I am sure that by now most people agree that the USDA categories are not totally informative. The public at large is especially concerned with the intensity and duration of any pain that research animals may have undergone in the course of experimentation, regardless of whether the animals received medication or not. Essentially, people want to know how much the animals really suffer. To provide that information accurately, the USDA categories would have to be changed or modified to resolve the public concern.

ALLEVIATING PAIN AND DISTRESS

It is important to note that the alleviation of pain and/or distress is not limited just to the use of drugs. In fact, the alleviation of pain and distress is often a very diverse task that includes environmental enrichment, the provision of nesting material and social enhancement in addition to or in lieu of drugs. I was pleased to learn that the proposed changes to Policy 11 actually might include nondrug treatment modalities under Column D of the USDA annual report. Presently, animals experiencing pain or distress and for which drugs have been withheld for scientific reasons are automatically placed in column E (unalleviated pain and/or distress) regardless of whether other treatment approaches were used. Of course, the appropriate categorization of experiments can be made only after appropriate veterinary monitoring of the study in conjunction with the investigator and the IACUC.

Laboratory animal veterinarians, animal care personnel, and investigators have a legal as well as moral obligation to alleviate pain and distress in animals. Alternatives to animal use in biomedical research should also be sought. Investigators will generally not use animals unless they have no other viable alternatives or unless they are unwilling to change the way they conduct their research. Animal research is expensive, it is much more difficult to control than in vitro experiments, and it is subject to a plethora of regulations. Reasonable investigators will try to avoid doing animal research if they can answer the same scientific question using in vitro methods instead. However, once the investigator has made a decision to use animals and the project is approved and funded, then the experiments must be done as humanely as possible.

TRAINING OF PERSONNEL

Among federal requirements is the provision that the IACUCs must assure the proper training of all personnel using laboratory animals. This mandate certainly includes the recognition and alleviation of pain and distress, and it applies to technicians, researchers, and veterinarians. To be sure, we cannot expect investigators or technicians to prescribe pain-relieving medications; this responsibility rests with the veterinarian. However, they should seek information and/or assistance when needed.

ROLES OF THE VETERINARIAN, IACUC, AND INVESTIGATOR

The AALAS position paper also emphasizes veterinary assistance of investigators, not only during the planning of potentially painful experiments, which is mandated by law, but also as soon as a study is under way. The statement also advocates adequate veterinary monitoring. Animal care and use protocols should be reviewed by the IACUC, and a pain category based on the expected pain level should be assigned at this time, that is, in a prospective manner. As soon as the study is funded and the investigator is ready to order animals, the veterinarian should arrange with the research laboratory to monitor or oversee the first few experiments or procedures in an attempt to ascertain the actual degree of pain or distress experienced by the animals. Whenever refinements are possible, the veterinarian should make pertinent recommendations to the researcher.

The results of this monitoring, as well as all associated recommendations, should be reported to the IACUC in a timely fashion. Based on this report, the IACUC should review the pain level category assigned originally and change it if necessary, in a retrospective manner, before submission of the USDA annual report. Therefore, more meaningful information can be forwarded to this regulatory agency. Involving the animal research staff in this process serves not only as an opportunity to train the laboratory personnel but also helps increase their awareness regarding the humane treatment of the animals. The primary monitoring responsibility may ultimately be delegated to the research laboratory, after the veterinarian and IACUC deem that its training is adequate. All observations and communications should be documented in a monitoring log. The veterinarian should continue to monitor these procedures, albeit more sporadically.

When designing monitoring schedules before the start of a new project the veterinarian should meet with the investigator and decide which pain- or distress-relieving medications could be used safely, without undue interference with the study. The responsibility for determining which drugs are appropriate rests with the investigator. Of course, the veterinarian must be familiar with the mode of action and pharmacological effects of these drugs so that he/she can make a valuable contribution toward this goal.

An important factor in reducing distress to the animals is their appropriate conditioning before a procedure. In other words, we should try to get them accustomed to the new procedure or the new environment in advance of the actual experiment. This acclimatization will also result in more reliable research data.

REFINEMENT

An important refinement that should be implemented whenever possible is the use of early endpoints, that is, ending the experiment at the earliest point that the desired data can be collected and before the animals are subjected to unnecessary pain and distress. Preferably, death should be avoided as an endpoint for animal experiments. Alternatives to death as an endpoint, such as behavioral changes, fluctuations in body temperature, HID_{50} (hypothermia-inducing dose 50 as opposed to LD_{50}) should be sought. Body condition scoring and weight loss patterns should be taken into consideration and evaluated to determine whether they can be implemented as a refinement. Pilot studies should be conducted under veterinary supervision whenever a novel procedure is planned that has the potential to be painful or distressful. For procedures that are known to be painful without the benefit of analgesics (e.g., sternal thoracotomy), pain-relieving drugs should be used preemptively.

CONCLUSION

As we all agree, to optimize the humane treatment of laboratory animals, more research is needed in the area of recognition and assessment of pain and distress.

QUESTIONS AND ANSWERS

DR. KARAS (Alicia Karas, Tufts University): With regard to your statement that analgesics and anesthetics could interfere with research results, I think the important point is that we understand quite well the effects of anesthetics and analgesics. However, the big unknown is how pain and distress affect research results. It appears to me that aspect of the issue should be included in the statement.

DR. COUTO: Your point is excellent. For the sake of clarity, this issue was addressed by the Scientific Advisory Committee and was included in the position statement. I did mention before that pain and distress, as well as the drugs used to alleviate them, can interfere. As a corollary to that, I should say that the effect of the drugs as well as the effects of the pain should be assessed or evaluated at least in a limited pilot study.

DR. ANDERSON (Lynn C. Anderson, Merck Research Laboratories): You said that animal reporting is the responsibility of the IACUC and the attending veterinarian, but you did not mention the investigator. I believe it is very important to stress that there is a real partnership between the investigator, the attending veterinarian, and the IACUC to monitor pain and distress; and that the burden is not on just the IACUC, the attending veterinarian, or the investigators, as suggested earlier.

DR. COUTO: You are correct. The burden for monitoring is shared, which is why we should involve research personnel in addition to the veterinarian and animal care technician. Sometimes when we can involve the investigator himself or herself, we should try to do that. Most often, however, principal investigators may not go to the animal room very often. It is easier to work with the contact person, a research associate, or the technician who is actually in contact with the animals.

On Regulating Pain and Distress

J. R. Haywood* and Molly Greene†

*Department of Pharmacology and †Institutional Animal Care Program
University of Texas Health Science Center
San Antonio, Tex.

I would like to share three messages with you today. First, the guiding principle for the development of all policies and regulations should be, "It must directly benefit the animals." Second, changes in pain and distress definitions, categories, and reporting should adhere to the dictum "Keep it simple." Finally, ethical issues in science are determined by advances in science; consequently, the more complex the science, the more difficult the ethical questions.

BENEFIT THE ANIMALS

When government policies and regulations are considered, the guiding principle should be that it must help the animals. Lack of adherence to this principle can be construed as contributing to regulatory burden.

What contributes to regulatory burden? One of the main sources of regulatory burden is the lack of understanding of the scientific process. A good example of this lack of understanding is represented by USDA's Policy 12. The goal of this policy is to provide guidance in enforcing the Animal Welfare Act, which requires investigators to minimize pain and distress by seeking alternatives to the use of animals in research. In the scientific process, the goal of the scientist is to test an hypothesis using the most rigorous scientific approaches possible. These approaches may require animals, cell cultures, mathematical models, and/or human subjects. The point is to do the best science possible. In reality, the scientist has already extensively considered alternatives to animals in designing the experiments. Good science requires more than a broad knowledge of the literature and a recent literature search. Unfortunately, there is a reality gap between the regulatory community and the scientific research community. Sessions like this workshop will contribute to closing that gap.

One of the suggestions from the study to help reduce regulatory burden was to prereview policies and regulations before implementation or commentary period. There should be discussion within the broad scientific community that includes investigators, IACUC coordinators, and laboratory animal veterinarians, all of whom will be affected by the policies or regulations.

In addition, there should be a sunset review of all policies and regulations that have been implemented. Every 5 or 10 years, policies and regulations should be reviewed by an external body to determine whether the policy has been effective and whether it should be revised.

The last aspect of what can contribute to regulatory burden is changing standards of compliance. In an environment where the rules keep changing, it becomes a constant challenge to remain in compliance, especially when there is no mechanism to directly inform investigators about the changes. Perhaps the rate of change should decrease so that everyone can understand what the standards are at this point in time.

If the factors I have described are the causes of regulatory burden, then what is the result of regulatory burden? What impact does regulatory burden have? Unfortunately, the impact is difficult to quantitate. However, issues such as lack of cooperation, feeling of distrust, and a general cynicism about the process can undermine the progress that has been made for animals.

Perhaps the most adverse effect of regulatory burden is the cost to the public. This cost is incurred with respect to financial loss due to reduced productivity as well as a loss of research. The burden of unnecessary regulations results in less research accomplished, fewer advances in health care, and fewer lives saved. Institutions and investigators have a shared responsibility with the regulatory agencies to minimize regulatory burden so that advances in health care can be made.

KEEP IT SIMPLE

The second message today is that any change in pain and distress definitions, categories, or reporting requirements must adhere to the dictum "keep it simple." Compliance is facilitated by keeping regulations and policy simple.

The Animal Welfare Act contains the following very simple statement: "The Secretary shall promulgate standards for animal care, treatment and practices in experimental procedures to insure that animal pain and distress are minimized, including adequate veterinary care with appropriate use of anesthetic, analgesic, or tranquilizing drugs or euthanasia." This simple statement has been made more complex than necessary.

The most effective means of minimizing pain and distress in research animals is not by federal regulation, but by an emphasis on the shared responsibility among investigators, veterinarians, laboratory animal care staff, and IACUCs to oversee the welfare of animals. There is an extensive process in place that, when operating effectively, ensures appropriate animal care. Pain and distress are considered in depth during the development of experimental protocols including veterinary prereview, IACUC review, and peer review by funding agencies and often by institutions. After protocols are approved and the work begins, animals are monitored throughout the study by animal care staff, research staff, and laboratory animal veterinarians. Consideration and observation by professionals in the institution are clearly the most effective means of minimizing pain and distress. In addition, this process is overseen by the IACUC during the semiannual inspection, the NIH assurance statement, the USDA in their annual inspection, and AAALAC International during the accreditation process.

With these assurances in place, why do we need to categorize experimental procedures and submit the numbers of animals in the respective categories to the USDA? As Barbara Rich of NABR pointed out at the SCAW meeting last month, the Animal Welfare Act only requires ". . . information on procedures likely to produce pain or distress in any animal and assurances demonstrating that the principal investigator considered alternatives to those procedures. . . ." I submit that quantitative information concerning animals experiencing pain and distress is not required by the Animal Welfare Act. In addition, because it fails to enhance the care and well-being of the animals, this requirement constitutes regulatory burden for the institutions. In their annual reports, institutions should be required to provide only qualitative information concerning the kinds of interventions performed on animals.

The question, "Do we need to change the classifications of pain and distress?" should be, "Do we need a classification system of pain and distress at all?" The answers are yes and no. No, we do not need one because the law does not require one, and the process of counting and reporting numbers of animals in different categories does not benefit the animals. Yes, we need a simple classification system to provide guidance and increase awareness of the investigators, laboratory animal veterinarians, and IACUCs. Such a system should be used only in the development, review, and implementation of a protocol to ensure that animals at risk receive extra attention.

The assessment of levels of pain and distress will always depend on individual interpretation. As Dr. Gebhart and Dr. Bayne told us this morning, the assessment will always depend on who is looking at the animals. This activity will be further complicated by the possible projection of personal feelings during the observations, which make it very difficult to categorize an animal regardless of how many subcategories are available. An additional complicating factor in the process is individual animal variation. As we heard earlier this morning, some animals recover from procedures better than others. This variation may relate to the differences in the individual sensitivity to pain, activity of the immune system, adaptation to stressful situations, or other factors we do not yet understand. The clear advantage of only reporting procedures is that the animal care team within an institution can focus on ensuring animal well-being instead of in which category an animal should be placed for reporting purposes.

ETHICAL ISSUES

My final message addresses the relation between biomedical research and the ethics of animal use. Jerrold Tannenbaum of the University of California, Davis, said at the PRIM&R meeting this spring, "Ethical issues in science are determined by advances in science." In other words, complex diseases and complex approaches to understanding disease often raise ethical questions pertaining to science. A recent example of this effect is the use of gene therapy to treat clinical problems. This exciting new technology, a spin-off of the human genome project, has developed rapidly. We are starting to see ways to treat some patients who could never before be treated. However, the pace of the science has exceeded the pace of the ethical considerations.

The use of animals in research has reached a similar juncture. Our biomedical research effort over the years has been too successful, and this success has created problems. People live longer in a more quickly moving society. As a result, problems related to aging, such as neurodegenerative diseases, cancer, and diabetes, occur more frequently. Concurrently, diseases of a modern, growing urban society, such as cardiovascular disease, mental health problems, chronic fatigue, and infectious diseases like AIDS, are more prevalent. Consequently, science has moved to the more difficult questions. The challenge to the research community has been to test scientific hypotheses related to the cause and treatment of these diseases with the best animal models that mimic these disease states while minimizing the animals' discomfort, distress, and pain.

Complex diseases require complex animal models. In response to this requirement, the scientific community has created disease models that could never have been developed without our rapid understanding of the genome. The explosive development of transgenic and gene knockout animals has opened up incredible new insights into previously difficult or impossible-to-study diseases. These new genetic models have been used with better pharmacological animal models, such as neurotoxin-treated animals to mimic Parkinson's and Alzheimer's diseases and surgically manipulated animals to replicate diseases like hypertension.

All of these approaches have permitted scientists to move to the next level of understanding, the cellular and molecular mechanisms of disease. In many laboratories, the use of animals has been replaced with alternative methods such as cell culture to study how signals are mediated within cells to change genetic expression or chemical interactions. An interesting dilemma is evolving. As we learn more about what happens inside a cell, it becomes more important to relate those findings to the whole animal. For example, the genome project is identifying hundreds of thousands of new genes that are responsible for the formation of a protein. After we learn what these proteins are, we will need to learn what they do, both at the cellular level and in the whole animal. In other words, the field of physiological genomics will likely require an even greater use of animals in research.

Again, it will be a shared responsibility in which the increasingly complex animal models of disease will mandate that investigators, veterinarians, and IACUCs have a greater responsibility ensuring animal well-being. Likewise, by helping to reduce regulatory burden and simplifying compliance with regulations in areas such as defining and reporting pain and distress, regulatory agencies will work with institutions to promote good animal care and good science.

QUESTIONS AND ANSWERS

DR. GLUCK (John Gluck, Kennedy Institute): I think your messages are very strong and important. You made the statement that regulatory burdens result in fewer advances in science. I have tried to document that premise in a number of different ways and have never been able to find any evidence. On what evidence is your statement based?

DR. HAYWOOD: I think it is Mark Twain who said, "So much conjecture on such a trifling of fact." I base that association on my observation that we are spending more money on compliance issues in universities. We are spending more money on indirect costs to support the research effort, including compliance issues. When that happens, less of the NIH budget is available for research. I use an exercise in logic rather than looking at any numbers because I think those numbers would be very difficult to capture. The cost of doing research has increased tremendously in this country. As a result, the amount of research that actually gets done on a dollar basis has been reduced. Perhaps Dr. Rich will address some of this in his talk later.

DR. RICH (Robert Rich, Emory University School of Medicine): You said it perfectly.

DR. HAYWOOD: This concern is a continuing one for FASEB; for example, whether enough money is going to research versus the support of research.

An Industrial Perspective

Lynn C. Anderson

Senior Director, Comparative Medicine
Merck Research Laboratories
Rahway, N.J.

Consideration of animal pain and distress is not a new issue. The utilitarian philosopher Jeremy Bentham addressed it almost 200 years ago when he wrote, "The question is not can they reason, nor can they talk, but can they suffer?"

For the purpose of today's workshop, however, I suggest that pain and distress be considered in the following framework:

1. The prevention and alleviation of pain and distress in laboratory animals is a legal and ethical imperative.

2. The quality of science depends on the absence of disease, injury, pain, or distress in our animal models.

3. The risk of causing pain or distress in a small number of laboratory animals may be justified if the research benefits society by identifying ways to prevent humans and other animals from suffering.

SOURCES OF PAIN

Pain can result from injury, surgery, or disease. It can be a potent source of stress and lead to distress or maladaptive behaviors; however, pain does not invariably lead to distress. Animals often adapt very effectively to pain or changes in their normal environment without becoming distressed.

As defined in *Recognition and Alleviation of Pain and Distress in Laboratory Animals* (NRC 1992), distress is an aversive state in which an animal is unable to adapt completely to stressors and shows maladaptive behaviors. This is the same definition of distress that Dr. Gebhart proposed this morning, and I recommend that USDA adopt it.

ASSESSMENT

The assessment of pain and distress depends on animal observation and professional judgment. Qualified personnel must use subjective measures to determine the presence of pain or distress. Attempts to categorize the intensity of pain or distress as "mild," "moderate," or "severe" would require even more subjective measures.

The assessment of pain and distress does not depend on any single indicator. A very broad spectrum of behavioral changes can be related to distress. The Carstens and Moberg (2000) article recently published in *ILAR Journal* underscores the importance and the diversity of behavioral responses.

We know there are significant differences in pain response between species, between breeds, within strains, and among individual animals. Their response is further modified by the animals' gender, environment, age, nutritional status, and weight. In fact, an infinite combination of factors can influence an animal's pain response. Therefore, it would be impossible to reduce the assessment of pain and distress to a prescribed checklist that would allow us to identify with certainty what is painful or not painful, distressful or not distressful.

SHARED RESPONSIBILITY

Institutions

Assessment of pain and distress should be a collaborative effort and include the investigator, attending veterinarian, IACUC, and animal care staff. However, the concern for pain and distress should begin at the institutional level. The institutional leadership must establish a culture that demands respect and consideration for laboratory animals. In addition, the institution is responsible for ensuring that the people who work with the animals are qualified. The responsibility for assessing personnel qualifications and training may be delegated to the IACUC or the attending veterinarian, but it is ultimately the institution that must set high standards for personnel qualifications and provide the resources for appropriate training. The institution is also responsible for ensuring that there is adequate veterinary and animal care, that appropriate equipment and facilities are in place, and that a fully functional and qualified IACUC exists.

Investigators

Investigators have the initial responsibility for assessing the potential for pain and distress in their laboratory animal models. They must consult with their veterinarians if there is the potential for pain and distress, but I advocate that investigators discuss their research protocol with a member of the veterinary staff to ensure that all aspects of the project support optimal animal welfare. Many of our investigators who are trained at the molecular or cellular level have very little experience with whole animal research. Often, the veterinarian is able to enhance the overall design of a research protocol, which surpasses regulatory requirements but supports good scientific method.

In consultation with the veterinarian, each of our investigators estimates the number or the percentage of animals that may experience pain and distress in each study. This information is provided in the protocol submitted for IACUC review and approval. If any animals are likely to experience pain or distress, the investigator must also provide a scientific rationale for proceeding with the study. Because we understand that prospective estimations of pain and distress will not be 100% accurate for 100% of the studies, especially when working with novel compounds, our IACUC requires investigators to notify the committee immediately if, during the conduct of their research, there are any deviations from the estimated number of animals that experience pain or distress. In addition, at the end of each government fiscal year, every investigator is required to report the actual number of animals that experienced pain or distress.

IACUC

As part of the protocol review process, the IACUC should evaluate the methods proposed to prevent or minimize pain and distress. The IACUC also has the responsibility to consider the scientific rationale for conducting a study that has the potential to cause pain or distress in laboratory animals and potential benefits to society. I agree with previous statements that anti-inflammatory agents and antibiotics, in addition to anesthetics, analgesics, tranquilizers and sedatives, are effective in relieving and preventing pain and distress. I support the USDA's proposed change that would include the use of all of these agents in Category D.

Veterinarians

A qualified veterinarian or animal technician should be involved in assessing the animals during the course of the research and provide pain relief whenever possible. More than 30 years ago, the AVMA adopted an oath that requires veterinarians to protect animal health and relieve animal suffering, which is very pertinent to today's discussion. However, veterinarians are also expected to use their scientific knowledge and skills for the benefit of society and to promote public health and the advancement of medical knowledge. We cannot forget that there are still many disease entities that we have yet to conquer and that this will depend on animal research. Therefore, in some studies, the veterinarian may need to consider the need to treat animal pain and distress along with the ultimate goals of the research and the potential benefits to both humans and animals.

PAIN AND DISTRESS INITIATIVES

HSUS

The HSUS pain and distress initiative suggests that we address the gaps in our knowledge of pain and distress and promote laboratory animal welfare by simply eliminating significant animal pain and distress. Although we understand many aspects of pain, more research is required to enable us to eliminate pain and distress further. We cannot address the gaps in knowledge by simply eliminating all research that results in significant animal pain and distress.

USDA

The USDA pain and distress initiative addresses the need to clarify the definitions of pain and distress and promotes expanding the categories for reporting pain and distress into mild, moderate, and severe. The biomedical community could benefit from additional guidance and clarification of the terms *pain* and *distress* by the USDA. However, there would be little, if any, added benefit to animal welfare by creating additional categories for reporting animal pain and distress.

I caution the USDA to avoid suggesting that certain procedures or animal models invariably induce pain or distress. For example, not all vascular procedures will cause pain or distress. Those that do will not be equal in intensity or duration of pain or distress. Likewise, not all animal models created by a specific technique, such as transgenic animal models, are subject to pain and distress. Any attempt to force certain procedures or models into specific pain or distress categories would preclude professional judgment. It would also fail to consider species-specific or individual animal variability. Finally, it would fail to consider the intensity or the duration of pain or distress.

Our investigators, veterinarians, and IACUC work diligently to evaluate prospectively whether a specific procedure in a specific research study may induce pain or distress. If we further expected them to estimate whether that pain would be mild, moderate, or severe, there would likely be differences of opinion. Although we may eventually come to a consensus regarding the categorization of pain or distress, similar professionals at another institution may have a different interpretation. This could lead to significant inconsistencies in the way animal pain and distress is reported and would not contribute to animal welfare.

REPORTING

Categorization

More complex reporting requirements will undoubtedly increase regulatory burden. It will certainly require more time and effort to categorize animal responses into different pain or distress categories that, in turn, will increase administrative costs. This regulatory burden does not benefit animal well-being nor does it benefit society; in fact, it may be detrimental to scientific and medical advances. It will decrease the scientific staff's available time for conducting their research. To put this issue in perspective, consider what would happen if 500 investigators each spent, conservatively, an extra 2 hours per year categorizing their animals' responses into mild, moderate, or severe pain or distress. That effort would require a total of 1000 hours. As a result, at least 6 months' worth of research work would not be done, which would likely delay or even impede important scientific discoveries. Animal welfare would not benefit; society may actually be denied improvements in health care or additional knowledge.

CONCLUSION

I would like to make several specific recommendations to the USDA. First, utilize the resources already available to assist investigators, attending veterinarians, and IACUC members to increase their understanding, recognition, and assessment of pain and distress. Specifically, I suggest informing them about credible references to use in this process. These include *Recognition and Alleviation of Pain and Distress in Laboratory Animals* (NRC 1992), the American College of Veterinary Anesthesiologists' position paper on the treatment of pain in animals (ACVA 1998), and the recently published Carstens and Moberg (2000) paper. I also encourage the USDA to use and enforce the existing methods available to them to ensure humane animal care and use. This responsibility includes ensuring the qualifications of research, support personnel, and functional IACUCs.

If the USDA makes any changes in the way we report animal use, they should devise a system to ensure that each animal is accounted for only once. We maintain a number of animals that are used from year to year in very benign studies; yet, the same animals are reported each year in Category C. If the intent of the annual USDA report is to identify how many research animals are used in the United States, then reporting one animal four times over a 4-year period is misleading at best.

Finally, I think the USDA should use the annual reporting mechanism to help educate Congress and the public about the value of animal research. Currently, the USDA annual reports do not demonstrate how the number of animals used have increased our basic understanding of biology and medicine or prevented pain and distress to other animals and humans by providing safe and effective therapeutic agents, medical devices, or surgical interventions. I urge the USDA to correlate their reports of research animals used to the ultimate benefits to society.

REFERENCES

ACVA [American College of Veterinary Anesthesiology]. 1998. Position paper on the treatment of pain in animals. JAVMA 213:628-630.

Carstens E., and G.P. Moberg. 2000. Recognizing pain and distress in laboratory animals. ILAR J 41:62-71.

NRC [National Research Council]. 1992. Recognition and Alleviation of Pain and Distress in Laboratory Animals. Washington, D.C.: National Academy Press.

QUESTIONS AND ANSWERS

DR. STEPHENS (Martin Stephens, Humane Society of the United States): Regarding your comments on the mild, moderate, and severe category system, you argued that there might be some inconsistencies in the reporting. However, the HSUS has documented fairly substantial inconsistencies in the current reporting scheme, and I believe that any inconsistencies in the new scheme should be weighed against that fact. I think over time, the inconsistencies in the new system would be ironed out after discussion at meetings like this, PRIM&R, and elsewhere. It could be that the same kinds of procedures should be classified differently depending on who is doing them and at what institution.

Finally, I would like to add that the current system seems to be as complex as any I have seen. In fact, throughout the 15 years I have been in this business, we have been trying to figure out what the current system means. I think it would be a little easier to tackle a system based on mild, moderate, and severe. Criticizing the new proposals and the direction we seem to be heading should be done in light of what we all agree is a very flawed current system.

DR. ANDERSON: I agree that there should be some clarification to ensure appropriate placement of animals in the existing C, D, or E categories. I think where we disagree is whether E needs to be further broken down into mild, moderate, and severe. I do not understand how those additional categories will benefit animals, how they will help us further prevent pain or distress.

Dr. Rowan said earlier that additional categories would help us focus on very severe cases of pain and distress. I maintain that we are already focusing on those cases, and we do not need to spend thousands of hours and hundreds of thousands of dollars in time and effort to get to those important issues. I do not see where mild, moderate, and severe pain categories are going to help us to do anything better for the animals.

MS. LISS (Cathy Liss, Animal Welfare Institute): I would like to add that I believe there is a serious problem of underreporting because of those who do not want to say there is any pain. If there is the opportunity to say it is mild, then it is more likely to be reported and, as a result, addressed by the IACUCs, the veterinarians, and others.

DR. HAYWOOD (J.R. Haywood, University of Texas Health Science Center): I am not sure animals that belong in Category E are underreported. I think the question is what these additional categories would do to help the animals. As Dr. Anderson just pointed out, animals that need attention are receiving attention, regardless of their classification.

MRS. STEVENS (Christine Stevens, Animal Welfare Institute): We need reporting, but some people are not reporting correctly. Some people do not care anything about whether animals suffer. A prime example currently is Dr. Frederick Coulston who, in his work with chimpanzees (obviously the most advanced animals that have to be watched) is unable to keep capable veterinarians because one after another have left. He has had young veterinarians directly out of school who have never had any experience with chimpanzees. As a result, 13 chimpanzees have died in the last year or 2, from horrible complications of not being watched.

I also want to draw your attention to the fact that veterinarians have different opinions. To say simply that veterinarians' judgment must be followed is flawed because you cannot proceed on the supposition that every veterinarian is equally able and equally humane. There are enormous differences.

DR. ANDERSON: There is a mechanism, through the USDA, to address animal welfare concerns at the Coulston Foundation. However, I would ask whether reporting animal pain and distress according to mild, moderate, or severe would make any difference in the quality of those animals' lives. Would checking a box to differentiate between mild, moderate, or severe make any difference in animal welfare?

MRS. STEVENS: I think it could make a difference because the National Institutes of Health are keeping Coulston from going bankrupt. This is taxpayers' money, and we are supporting him by the tens of millions of dollars. I object to that money very personally.

DR. SCHULTHEISS (Peter Schultheiss, USDA): As a laboratory animal veterinarian who has spent the past year working with the USDA in animal care, I would like to address Dr. Gebhart's concerns regarding the value of reporting (not of breaking down "E" into mild, moderate, and severe). In reviewing the annual USDA reports, I learned something of which I was not aware via any other mechanism: that according to the statistics, about 80% of all animals in Category E were involved in testing, and more than half of those animals were used specifically for biologics testing for vaccine potency, purity, and safety.

While I was with the USDA, I received a call from Dr. Stokes, who asked me for a relative breakdown of testing procedures so that we might be able to concentrate our efforts on developing alternatives where they are most needed. Given the way we collect information now, I was not really able to pinpoint it, but I was able to bracket animal numbers as to what effect a certain alternative would have. That use of data would be an example, I think, of the value of collecting information for targeting where alternative work is needed.

I have a separate question for Dr. Anderson concerning her statement that the present system fails to consider intensity or duration. Other than pain or distress that is not more than slight or momentary, you are correct, there is not really any duration or intensity consideration. I would ask you for a proposal to include that issue in a way that would not complicate the system so much that reporting really would become a burden. I think it would be a great idea to incorporate it according to agreed breakdowns such as 5 minutes, 10 minutes, 2 days, 2 weeks, or whatever is appropriate.

DR. ANDERSON: My point was that the USDA should not attempt to place certain procedures or animal models into specific pain and distress categories because animals do not respond to the same procedures with the same intensity or duration of pain and distress. For example, ovariohysterectomy may induce a very different pain response in a mouse than it would in a dog or other species. In this example, the pain or distress associated with the procedure would depend on many factors, including the interventions used to prevent pain. If, for example, the USDA were to suggest that all abdominal procedures cause severe pain and distress, the use of professional judgment would be precluded.

DR. GLUCK (John Gluck, Kennedy Institute): With respect to the point about how is it going to benefit animals to categorize pain, I think the answer is in your own talk. You started with Jeremy Bentham and the ethical system that he promulgated, which of course requires a judgment of right behavior based on consequences. I think if I, as a researcher, am faced with having to confront both my own and my expert colleague's judgment about the consequences of a particular intervention, whether it produces extreme, moderate, or mild pain, this ethical system will permit me to make explicit judgments about the ethical basis of my own experimental work. I believe it is valuable in that previous speakers have talked about our incredible ability as human beings to distance ourselves from some of the consequences of what we do, especially when we know our heart is really in it and we are really interested in those kinds of issues. I think the animals potentially can benefit, and I think the science can also benefit.

Corners Still Unswept

John E. Harkness

Mississippi State University
Mississippi State, Miss.

With my rabbit and rodent talk only 2 weeks in the making and with no prior experience among the usual circuit riders of distress and pain, I could think only of introduction after introduction with only modest attempts at discussion and conclusions. So I will give you my four speech introductions and make a stab at the rest.

INTRODUCTIONS

My Life as a Rat

When I was 7 years old, my intelligent and sensitive linoleum-layer father took me and my brothers for 4th of July fireworks at Edgewater Park on the Rouge River in Detroit. The sun set, slowly, and then the noxious, exploding stimuli began. My nociceptive apparatus activated my physiological and behavioral responses through those rapidly receptive limbic and isocortical parts, especially my amygdala and somatosensory cortex. Then began my perceptions, whose true dimensions were known to me only. The explosions were directly above me. The aerial bombs hurt my ears, so my fingers went into my ear canals. I vocalized, "get me out of here." I wept. I cringed. I was anxious, fearful, and nauseated. I wanted to escape. I was stressed, in distress, maybe agony. I was suffering and depressed. My well-being, quality of life, welfare, happiness, contentment, life satisfaction, comfort, pleasantness, positive feelings, sense of control, and freedom from frustration, helplessness, and hopelessness were going fast. Even my endogenous opioids failed me. I cannot quantify or define the feelings well, but the sum total was both painful and in retrospect distressing.

There were no child welfare inspectors to be seen, no one around concerned about my experiments in maturation, and no one, including my father, seemed to notice or to care, except my father called me a sissy and told me to stay put and keep quiet. I did. I survived, shaking. But like the residents of Vicksburg, Mississippi, I still have limbic, neuroendocrine, and autonomic reactions on the fourth of July, when we both lost a battle.

Remembering my emotions and pain that night in the late 1940s, I recently asked my neurologist colleague whether the nociceptive anatomy of rodents and rabbits resembled mine. His initial answer was rapid, emphatic, and intuitive: "I believe placental mammals, despite some differences, due to modifiers of the responses, are very similar." Period. I learned also from three decades of books and articles that primate nociceptors, A-delta and C sensory fibers, spinal tracts, brain tracts, the limbic system, thalamus, somatosensory cortex, and brain/weight ratio seemed to be very, very similar to those in rats, even more alike than between those beloved cats and us human primates. I was comforted also to learn that most aversive or avoidance behaviors required no isocortex; my prereptilian brain parts could protect or harm me, rodents, and rabbits alike, quite well on its own, without much isocortical or frontal lobe override.

Then, recently, I wanted to know why my father had missed my obvious signs. After all, he had been a kid once. I wanted to know whether proxy assessment by humans of distress and pain in other talking humans, human infants, and in placental mammals was valid, reliable, and sensitive.

Lynne Holton and colleagues (1995, p. 64) of Glasgow and Edinburgh described in JAVMA that "currently, there are no effective objective methods of measuring the intensity of clinical pain." No mention of distress. Dr. Horton continued to state that one must rely on subjective assessments that must be valid, reliable, and sensitive. She favored having trained teams use a numeric scale for pain evaluation, but that scheme had deficiencies also. Francis J. Keefe and others (1991) at Duke and elsewhere published in *ILAR News* that well-trained observers using an observation scale with talking children found high correlations between the two groups' interpretations of the test group's behaviors to stressors. I would hope so. Franklin D. McMillan (2000) stated in JAVMA that "Numerous studies provide evidence to suggest that proxy responses by parents correlate poorly with the perceptions of the child they are representing—especially on certain subjective feelings regarding illness and emotional states" (p. 1908). He believes there is no valid, current instrument for measuring QOL in animals. He advocates assembly of a diverse population of experts to establish measuring instruments and conduct research on their use in animal studies. So do I.

Yet, despite these questions of variability among individuals and groups in linking overt behaviors with degrees of distress and pain, several excellent sources, including those published by J. Wallace Fiat (Convenor and others) in 1990 in *Laboratory Animals*; in *Recognition and Alleviation of Pain and Distress in Laboratory Animals* (NRC 1992); and in Carstens and Moberg's (2000)

"Recognizing Pain and Distress in Laboratory Animals" in *ILAR Journal*. These publications listed acceptable, realistic, believable signs of pain and even distress in rodents and rabbits. Those lists are of critical importance and should be assembled, reviewed, and known much more widely by those who use rabbits and rodents in teaching and research. The challenge is to inculcate into our institutions those signs and their significance. Some investigators do not always listen or read their mail. My experiences in 1947 in Edgewater Park continue in our animal facilities.

My Ignorant Colleagues

I alone at this session, with the possible exception of Dr. Karas from Tufts, am involved full time working as a Johnny-on-the-spot, university-wide laboratory animal veterinarian with mostly veterinarians as investigators. Despite how you react to Trent Lott, the movies about Mississippi, or the many Connie Chung-Dan Rather magazine shows about Mississippi, our college of veterinary medicine has a fine facility, faculty, and staff. We are fully accredited by everybody interested, including AAALAC. The research investigators and active coinvestigators often have veterinary and PhD degrees, private and institutional practice experience, and speciality boards. They do not see distress and pain as I do, however, and both in 1994 and in 1997 our surgeons and anesthesiologists argued vigorously with the AAALAC visitors over valid signs of pain (no one mentioned distress) and what to do, if anything, with analgesics. Non-pain induced distress was never mentioned. Their position was to use "when needed" or "professional judgment," popular catch words in our domain, to determine when pain relief was appropriate and when it was not. The revised USDA Policy 3, especially the section on records, which I have placed under their noses at least three times, does not impress them; only intensive IACUC oversight of protocols and procedures, very time consuming, gets their attention. Now the HSUS draft proposal comes along and tells me that even our IACUC's and my rather strict placement of procedures into the usual pain categories nowhere meets the level of recognition of emotional distress and perceived pain that is standard in Europe. I wonder how we would rank with the norms in the advanced cultures of Asia, Africa, Southern Europe, and Latin America.

One of our best clinical researchers, with a PhD in pharmacology from the University of Illinois, e-mailed me last week saying, "John, enough is enough" when I told him to refine and describe his potentially painful surgical procedures on dogs that would not recover from anesthesia. He told me to do it; he apologized later. He remains a believer in the outdated concept of a vertical phylogenetic scale. Even our experienced 15-person IACUC is refractory to expanded legal promulgations, and instead the committee champions professional judgment because they have been given that encouragement on many occasions. I myself can still manage to influence the committee and meet the mark, almost, because I experienced similar resistance in the 1970s and 1980s and became accustomed to the recalcitrant. But I am getting tired and hard to replace. My newly hired, very competent, young clinical laboratory animal veterinarian places compliance assessment low on her list of concerns and future rolls. I will continue to handle that aspect until some compliance officer with an MBA can take over when I retire in 5 years.

My colleagues do believe, however, what I read often: There are many tried and true methods for the reduction of distress and pain in animals if drugs and nondrug procedures used properly are study compatible and if considerable effort is made to define in clear, legal language the terms distress, pain, and perhaps some other, related terms.

Rats and Your Cousins, I Love You

In my 17 years, 8 on the AAALAC Council, as a site visitor and also as a part-time consultant to small Mississippi colleges with domestic rat and mice colonies only, I have seen hundreds to thousands of rodents and rabbits in their barren, sanitized twice a week plastic or metal or wire cages, on treadmills, abdomens open on surgery tables, burned by infrared light, injected with turpentine, living but with skin burned in a -17° F freezer, operated on without shaving, having their skulls drilled and spinning on rotating bits, carrying ulcerated tumors, all with no USDA oversight, little if any IACUC concern, scant environmental enrichment, and no perioperative distress and pain relief, drugs or otherwise. I just reviewed two "big journal" manuscripts for studies involving surgery in rodents in which there was no description for relieving distress and pain, not even "as needed" in someone's eyes. Presumably their IACUCs approved of this. I did not. There are certainly many more studies I could describe, but with my memory more limbic than cerebral, I remember no more. I only wish I had time to tell you how complex is the behavior of rabbits and rodents; their brains are marvelous organs, and they must perceive more than we really want to know. For example, each mouse whisker has a special sensory section processing in the somatosensory cortex, which gives consensual barbering among cagemates a new dimension.

I do not know how many domestic rats and mice are used in research, teaching, or testing each year, few people seem to know exactly, nor do I know into which pain categories they fit or should fit; all I hear is that mouse use is "exploding." But the common mouse-rat usage is usually stated to be between 18 to 22 million with each animal with nociceptive structures and responses to distress and pain very much like ours. All this possible distress; and so little known outside of locked IACUC file drawers. For regulated rodent relatives and for rabbits used in 1998, some numbers are known: 160,000 guinea pigs; 200,000 hamsters; and 285,000 rabbits. Of those, 100,000 guinea pigs, 100,000 hamsters, and 135,000 rabbits were reported in pain categories D or E: 335,000 of 745,000 total—45% in categories D or E. In my opinion, more should have been placed in Category E. But 45% compares well with reports in the Netherlands. But these animals' populations in research are declining rapidly, unlike those of mice.

Overwhelmed

When Ralph Dell asked me 3 weeks ago to talk, I asked our librarians to do a 1995-2000 Medline and Agricola literature search on key words: rats-pain, mice-pain, and rabbits-pain. In a few hours, Mrs. Grimes called to say that Medline had 3000 rat-pain articles and 2000 mouse pain articles, and Agricola had many also. So much information, so little coordination, so few journal reviews.

CONCLUSIONS

1. Someone out there must think, are thinking, about distress and pain in rodents and rabbits and preaching convincingly. Yet, the open positions and salaries for persons willing to do what I do seem to be expanding while the number of available, qualified applicants with clinical training is declining. Even NIH does not train John Harkness types any more.

2. Establish definitions of distress and pain, using the ones we have as a basis. In all the papers and books I have read, I have seen many very understandable, useful definitions. How much more time needs to be spent? But if you flood us with other subjective, unmeasurable terms and flow charts, we will laugh and lose interest. Speak to us as does the *New York Times*, as busy 12 year olds, not as desk-bound philosophers.

3. Promulgate again and again in palatable form with one voice the signs of pain and non-pain induced distress. We accept those oft published lists for rabbits and rodents. We know the signs. Now we need on that subject more communicators and fewer committees.

4. Develop valid, reliable, sensitive and model scales for groups to evaluate and rank our perceptions of an animal's perceptions. Publish the research results and keep the language simple.

5. After doing the above, or after promising to revise whatever you do prematurely, give us more detailed ways to categorize distress and pain, even if the purpose is to inform the IACUCs of the nature of distress and pain of the study under review. You will find me and our IACUC very receptive.

In addition, I think in this country only the *Guide* and the USDA regulations, but regulations above all, can accomplish these goals, if the authors are sufficiently research wise and are willing to compromise on details, reduce debate, and promulgate practical and reasoned doctrines.

REFERENCES

Carstens E., and G.P. Moberg. 2000. Recognizing pain and distress in laboratory animals. ILAR J 41:62-71.

Fiat J.W., J. Sanford, and M.W. Smith. 1990. The assessment and control of the severity of scientific procedures on laboratory animals. Lab Anim 24:97-130.

Holton L., E.M. Scott, A.M. Nolan, J. Reid, E. Welsh, and D. Flaherty. 1995. Comparison of three methods used for assessment of pain in dogs. JAVMA 212:61-66.

Keefe F.J., R.B. Fillingim, and D.A. Williams. 1991. Non-verbal measures in animals and humans. ILAR News 33:3-12.

McMillan F.D. 2000. Quality of life in animals. JAVMA 216:1904-1910.

NRC [National Research Council]. 1992. Recognition and Alleviation of Pain and Distress in Laboratory Animals. Washington, D.C.: National Academy Press.

Personal Experiences with Clinical Pain Management, Study Design, Mitigation of Scientific Confounders, and Long-term Gains to the Researchers and Public

Victoria Hampshire

Advanced Veterinary Applications
Bethesda, Md.

It is truly an honor to be invited here to the National Academy of Sciences and to have the opportunity to be associated with all of you. As a newcomer to this arena, you might say that I am well qualified and have come of age on the front line in mammoth research programs where I have collaborated with scientists to reduce distress in animal models within scientific constraints of the protocols.

EXPERIMENTAL EFFECTS ON ANIMALS

Animal models of human disease and physiology are becoming exceedingly complex. Experimentation is not neatly packaged. Pain and "altered states" of physiology leading to distress can be acute or chronic in duration and any combination. It is a mistake to state, as many have, that animals do not suffer during experimentation simply because they have not been observed to suffer. Lack of observation is particularly relevant because only small numbers of programs, if any, provide more than 8 hours of care over the workday. It is also a mistake to dwell on costs associated with increased provisions for monitoring and intervention programs because the relative savings are well known in terms of higher animal yields, smaller interanimal variability due to management of stressors, and shorter time from bench-to-bedside human trials. Thus, quality pain management programs result in more observations during experimentation and public assurances that are immeasurable.

Results of Cumulative Minor Events

It is my clinical opinion that most animals in research suffer from cumulative and minor events that, when combined, amount to distress. In other words, the animal is not feeling well enough to normally ambulate, eat, or drink. It then becomes a bit dehydrated, a downward trend develops, and more serious stressors result. Thus, early detection provides the greatest gain in terms of control of variables and pain and distress management. Current endpoints such as weight loss and low body temperature are instituted long after distress and stress or pain are encountered and are useless to refinements that are either pragmatic or beneficial. Viewed prospectively, however, a variable such as weight can be very useful for finding that particular instance when results become negative. Time is therefore essential in pain management if one hopes to reach beyond a paper program and create one of substance. The importance of time is more dramatic in smaller animals as metabolic rate is known to be roughly 10 times higher in mice than man. Other species fall somewhere between. The best way to describe this early detection of weight loss and other variables contributing to physiological stress is to describe the following example.

Multidimensional Risk

An excellent example of multidimensional risk is that of canine myocardial ischemia (Banai and others 1991, 1994; Lazarous and others 1996; Rajanayagam and others 2000; Shou and others 1997; Unger and others 1990, 1991, 1993a,b; 1994). The dog patient received a left lateral thoracotomy incision to create a left anterior descending coronary artery event. In this example, I shall discuss veterinary support surrounding a decade of research involving hundreds of canine patients on which human trials were eventually predicated. The instrumentation utilized to constrict arteries over a 3-week period of time is a silastic balloon ameroid placed over the artery and tunneled through the ribs into the subcutaneous space. The dog, pig, or rat (as are today's models) is then recovered overnight with oxygen, warmth, analgesics, and antibiotics; it also must receive constant monitoring for ventricular arrythmias, infection, electrolyte imbalance, and glucose disturbances. The model then undergoes the additional stress of serial MRI or angiographic episodes under general anesthesia. Then add the additional experiment whereby the investigator wishes to administer viral-mediated gene therapy using adenoviral vascular endothelial growth factor, and have a singular experiment within a multidimensional risk project. The risks to distress include infection, dehydration, electrolyte disturbances, arrythmias and angina, poor wound healing, and increased catabolic demand at a time when animals do not feel particularly well.

Within this one model, you may look at one possible risk such as infection or septicemia. Septic animals undergo an initial systemic inflammatory response.

The body is then riddled with the events associated with hyperdynamic shock, leaky blood vessels, pulmonary edema, and glucose and electrolyte disturbances. Such events must be detected within hours to provide symptomatic relief as well as to stabilize the experiment. Therefore, an 8-hour or singular monitoring scheme is worthless to the animal, the model, and, ultimately, extrapolation to public medicine and benefit.

PROVISION OF NECESSARY IMPROVEMENT

In 1989, when I first joined NIH, this model received only a few hours of intensive care and was put back in a kennel. The mortality rate was 55%. During my first 5 years there, I had supportive program directors who encouraged the augmentation or magnification of veterinary presence. I made steady improvements, which included acquisition of an overnight technical staff, clinical chemistry and complete blood count analyzers that gave results instantly at the cageside, and the provision of scoring systems that augmented analgesic administration. Survival rates increased to 95%. Deaths were always associated with sudden ischemia and closure of the ameroid rather than with other complications.

I was very proud of this progress and other changes, and the scientists noted a more expensive short-term solution with long-term benefit. Over the next 5 years, I continued to make similar improvements across all species and all projects. Although the scientists initially viewed this as expensive, they eventually understood the benefit.

Scheme of Veterinary Care

In my view, the way to achieve this outcome is to suggest a scheme more like a good veterinary teaching hospital or private clinic for animals in these risk groups. We thus need small teams to cover large amounts of ground and high rodent density housing for less risky groups in an effort to discover outliers. Such management is accomplished by the hiring of clinically astute veterinarians and roaming technical teams.

The emergence of large numbers of genetically altered rodents can also be monitored in this manner by central dispersion of teams of technologists under the line command of clinical veterinarians. Successful monitoring has already been achieved in some places and was recently described in *Lab Animal* (Hampshire and others 2000a,b).

Shifting Responsibility for Performance Standards

Additionally, it is not reasonable to expect today's scientist to be clinically knowledgeable or experienced about veterinary medicine; therefore the development and line accountability of such teams of clinical veterinarians and technologists are absolute requirements for accomplishing this task. In the context of research budgets, this approach is easily achieved by a shift from engineering standards toward greater emphasis on veterinary performance standards. However responsible, the scientists can be unrealistically expected to understand or have time to fully manage such key animal populations. There must be responsibility, collaboration, and authority of clinical veterinary staff.

ESTABLISHMENT OF PREREVIEW

Finally, part of ILAR's directive today was to provide guidance to USDA's inclusion of alternative searches in Policy 11. I have not heard anyone call this a similarities search, but having performed such searches, I believe the existing directive does not enhance science or animal care. The current policy and enactment drive away scientists rather than enlisting their cooperation because it is described and viewed as a barrier rather than an opportunity. A better solution is to require a prereview in which the attending veterinary staff search human and animal-similar literature for the purposes of seeking answers to confounding variables, stress and pain outcomes, and case management of human or animal-similar patients.

Alternatives, in my opinion, are most beneficial when viewed as refinements to animal care and study design. If such a search were undertaken prospectively for these purposes, compliance with responsible use of animals would often develop naturally.

SCIENTIFIC JOURNALS AS FACILITATORS

I must mention literature because methods that describe adequate animal care and monitoring are frequently not part of scientific journal reporting. Many investigators argue that their science does not need refinement because the methods they are utilizing have been reproduced time and again according to proven methods. Many will also try to pool controls from previous work in which no pain and distress management schemes were utilized. I contend, however, that many methods are missing from such papers. A preponderance of papers reviewed, do not mention analgesic programs and leave the reader to wonder if compliance really existed at all.

One must understand that the recalcitrant scientist will continue to ignore Policy 11 directives but will still cater to the scientific journals under today's "publish or perish" conduct. If the majority of exemplary publications were also to include a section specifically describing pain and distress monitoring, duration, and intervention criteria, the mainstay of scientists would also conform to that standard. A commensurate education in Policy 11 guidelines and journal management would be very useful for achieving this goal.

CONCLUSION

I have attempted to describe a pain classification system that is substantive when viewed retrospectively. It assumes strong veterinary action in prereview, the design of a pilot, and retrospective adjustment of protocols not only so that accurate reporting is performed, but also so that the public can be made more aware of which relief measures were meaningful. This system of veterinary collaboration, pilot design, and retrospective refinements and reporting affords more efficient experimental conduct with more accurate reporting of results and animal pain classification.

REFERENCES

Banai S., M.T. Jaklitsch, W. Casscells, M. Shou, S. Shrivastav, R. Correa, S.E. Epstein, and E.F. Unger. 1991. Effects of acidic fibroblast growth factor on normal and ischemic myocardium. Circ Res 69:76-85.

Banai S., M.T. Jaklitsch, M. Shou, D.F. Lazarous, M. Scheinowitz, S. Biro, S.E. Epstein, and E.F. Unger. 1994. Angiogenic-induced enhancement of collateral blood flow to ischemic myocardium by vascular endothelial growth factor in dogs. Circulation 89:2183-2189.

Hampshire V.A., C. McNickle, and J.A. Davis. 2000a. Red-carpet rodent care: Making the most of dollars and sense in the animal facility. Lab Anim 29:40-45.

Hampshire V.A., C. McNickle, and J.A. Davis. 2000b. Technical team approaches to rodent care: Cost savings, reduced risk, and improved stewardship. Lab Anim 29:35-39.

Lazarous D.F., M. Shou, M. Scheinowitz, E. Hodge, V. Thirumurti, A.N. Kitsiou, J.A. Stiber, A.D. Lobo, S. Hunsberger, E. Guetta, S.E. Epstein, and E.F. Unger. 1996. Comparative effects of basic fibroblast growth factor and vascular endothelial growth factor on coronary collateral development and the arterial response to injury. Circulation 94:1074-1082.

Rajanayagam M.A., M. Shou, V. Thirumurti, D.F. Lazarous, A.A. Quyyumi, L. Goncalves, J. Stiber, S.E. Epstein, and E.F. Unger. 2000. Intracoronary basic fibroblast growth factor enhances myocardial collateral perfusion in dogs. J Am Coll Cardiol 35:519-526.

Shou M., V. Thirumurti, S. Rajanayagam, D.F. Lazarous, E. Hodge, J.A. Stiber, M. Pettiford, E. Elliott, S.M. Shah, and E.F. Unger. 1997. Effect of basic fibroblast growth factor on myocardial angiogenesis in dogs with mature collateral vessels. J Am Coll Cardiol 29:1102-1106.

Unger E.F., S. Banai, M. Shou, M.T. Jaklitsch, E. Hodge, R. Correa, M. Jaye, and S.E. Epstein. 1993a. A model to assess interventions to improve collateral blood flow: Continuous administration of agents into the left coronary artery in dogs. Cardiovasc Res 27:785-791.

Unger E.F., S. Banai, M. Shou, D.F. Lazarous, M.T. Jaklitsch, M. Scheinowitz, R. Correa, C. Klingbeil, and S.E. Epstein. 1994. Basic fibroblast growth factor enhances myocardial collateral flow in a canine model. Am J Physiol 299(Pt 2):H1588-H1595.

Unger E.F., C.D. Sheffield, and S.E. Epstein. 1990. Creation of anastomoses between an extracardiac artery and the coronary circulation. Proof that myocardial angiogenesis occurs and can provide nutritional blood flow to the myocardium. Circulation 82:1449-1466.

Unger E.F., C.D. Sheffield, and S.E. Epstein. 1991. Heparin promotes the formation of extracardiac to coronary anastomoses in a canine model. Am J Physiol 260(Pt 2):H1625-H1634.

Unger E.F., M. Shou, C.D. Sheffield, E. Hodge, M. Jaye, and S.E. Epstein. 1996. Extracardiac to coronary anastomoses support regional left ventricular function in dogs. Am J Physiol 264(Pt 2):H1567-1574.

Use of Laboratory Animals in the Postgenome Era

Robert R. Rich

Emory University School of Medicine
Atlanta, Ga.

I am pleased to have the opportunity to bring the somewhat removed but, I hope, useful perspective of a dean to the important questions being addressed today. I would like to begin with several general observations on the interface between new trends in biomedical research and their impact on the issues of pain and distress in laboratory animals. These observations require the caveat that my perspective is from a school of medicine. Explicitly, some of my observations and generalizations do not apply to other entities, such as schools of veterinary medicine, because I will discuss biomedical research as it relates to human disease.

PREDICTABLE TRENDS IN RESEARCH

Consider with me how completion of genome sequencing in humans and numerous model species will likely change biomedical research over the next 20 years, a time frame I will refer to as postgenome medicine.

Systems Biology

I assert that we will see the ascendancy of what I would call systems biology. We have heard several times today that in the past decade, the interest in experimental animals has increased as molecular biologists, utilizing the power of transgenic and gene knockout animals, have moved their focus from the molecular laboratory to the intact animal. I am confident that over the next 20 years, this trend will accelerate. Additionally, this trend will require scientists with molecular training to focus increasingly on the complex interactions between cells and tissues and whole organ systems that can be accomplished only in living animals. Consequently, there will be a marked increase in use of and dependence on research involving animal models. Thus, we should recognize that there will not be a reduction in the total number of laboratory animals used. On the contrary, to take full advantage of the postgenome revolution in biomedical research, we will increasingly call on the use of animal models to solve problems of human illness.

Molecular Genetic Technology

Nevertheless, I also believe there will be a considerable reduction in the heterogeneity of animal models and species used. We will focus instead on those animal models with which we can most effectively exploit the power of molecular genetic technologies. We will observe a particularly dramatic increase in the use of rodents as novel animal models of human diseases. In contrast, at schools of medicine (and I explicitly exclude schools of veterinary medicine) I believe the use of most large animal species will decline substantially. Of course, there will be some important exceptions to this latter generalization. For example, new advances in xenotransplantation will require certain large animal models that will be uniquely useful because they have been genetically engineered for experimental, and possibly within 20 years clinical, use. Similarly, nonhuman primates will remain uniquely useful for certain purposes, such as studies in cognitive neuroscience.

I believe one of the more important consequences of the availability of new molecular technologies is that they will change the very nature of the use of laboratory animals. This use will result in new experimental capabilities. Increasingly, the basis of pain and distress in laboratory animals will not be a reflection of something done to them other than the alteration of their genes. Laboratory animals, predominantly mice, fish, and invertebrates, will be genetically manipulated in ways that result in development of disease or functional disorders. In that sense, they will resemble human beings who are genetically predisposed to different diseases. The animals will develop these disorders or diseases as a consequence of genetic alterations of the germ line.

PREDICTABLE CONSEQUENCES OF GENETIC BREEDING

I believe that our understanding of novel approaches to the treatment and prevention of human (and animal) diseases will be greater than we can even imagine today. As a corollary, however, I also believe that it is unrealistic to have as a foreseeable goal the elimination of pain and distress in experimental animal models that replicate or mimic human diseases. I submit that as long as we can develop animal models by genetic manipulation and gain novel insights into cures, preventions, and treatments of diseases, we should focus on the management rather than the elimination of pain and distress.

Definitions and Categories

We have heard considerable discussion today about the current USDA definition and protocol categorization based on anticipated pain and distress. I agree with speakers who have argued that the current definitions are flawed. I believe part of the flaw is the focus on process rather than outcome. For example, the categorization includes such variables as the use of analgesia or anesthesia, rather than simply asking the more simple (and important) question of whether a proposed protocol would involve an element of animal pain and distress. In this respect, I believe the approach of the Humane Society of the United States, simply to classify pain and distress according to three categories of outcome, represents an improvement. It is important to have a process whereby investigators, veterinarians, and IACUCs assess protocols and monitor research in an effort to minimize pain and distress.

Dr. Haywood admonished us earlier today to keep it simple. I agree completely. Laboratory animals are best served by a simple classification that speaks exclusively to the outcome. We need to use simple words that all of us generally understand and can apply in global assessment, as opposed to overdefining or overquantifying such variables as temporal and intensity issues. I believe a more simple clinical judgment would focus more appropriately and consistently on qualitatively minimizing animal pain and distress and would use the practical experience and intuitive judgment of trained veterinarians and investigators. In fact, I am not truly convinced that it is useful to have distinct definitions of pain and distress. We know what those words mean. The issue is simply whether they are absent (or minimal), moderate, or severe.

Procedures

I believe that exemplary procedures generally associated with moderate or severe pain are useful as guides to investigators and to IACUCs. I would argue, however, that such exemplars should not be codified in regulation or policy, which risks replacing good judgment with a "check-off" mentality. I believe the more simple notion of global assessment, as I interpret the Humane Society categorization, serves animals well and provides a regulatory framework that can be applied and enforced with greater consistency. Simplifying the process of protocol preparation and review, and elimination (to the extent possible) of the "hassle factors" that promote investigator cynicism and reluctant cooperation, would also contribute significantly to realization of our shared goal of minimizing pain and distress.

Replication of Human Diseases

Nevertheless, I want to reemphasize that just as human disease inherently involves pain and distress, it will not be possible to eliminate these conditions in laboratory animal research as we attempt to replicate human diseases. However, one of the attractive aspects of the Humane Society's approach for simplifying the categorization of pain and distress is to facilitate work among investigators, veterinary staff, and IACUC members to try to "down-categorize" otherwise painful or distressful protocols with appropriate use of drugs and other modes of pain and distress relief. We should focus on the overall pain or distress experienced and incorporate modalities of pain or distress relief into study designs. Perhaps then we will be increasingly able to classify protocols not as Category E but instead as Category D or C because the appropriate interventions have been applied.

Experimental Endpoints

I also advocate increased attention to and definition of specific experimental endpoints that can minimize pain and distress. This would be another advantage of a more simple categorization that, instead of emphasizing the processes involved, requires an answer to the following questions:

"Did the animal utilized in this study experience moderate or severe pain or distress?"
"If so, could we have down-categorized an otherwise painful procedure by choosing an earlier experimental endpoint?"

Obviously, earlier experimental endpoints might be as effective as interventions with drugs.

CONCLUSION

Finally, I want to echo the other speakers who have pointed out that a cooperative venture is required. No one involved in the responsible conduct of science does *not* want to minimize pain and distress of laboratory animals. High-quality science requires that we do this. The investigators who design and execute a study, the IACUC members who review the protocols, and the veterinary staff who monitor the process all share the goal of humane care and treatment of our incredibly valuable laboratory animals. By working together, I believe we can develop policies and procedures that are relatively simple, can be understood, can be consistently applied, and are, in fact, good for the animals.

QUESTIONS AND ANSWERS

DR. DE HAVEN (Ron DeHaven, USDA): Dr. Rich, you suggested that we not codify examples of the categories (minor, moderate, and severe) in regulation or policy. How then, from a regulator's standpoint, would you ensure uniform interpretation of categories?

DR. RICH: Your question is excellent. I do believe that exemplars are helpful. I suggest that we, as a community, can agree on some exemplars but still avoid encouraging checklist mentality inspections of university animal usage, which lack a thoughtful approach to what was actually done. I believe we are ill-served when checklists and clipboards replace good judgment. However, having some community consensus regarding certain procedures would ensure that appropriate corrective measures would be taken if an inspector found substantial discordance from this consensus in a particular institution's categories of procedures or protocols. However, to the extent that codification of exemplars would tend to focus on process rather than outcomes, I believe this approach could actually make inspection and enforcement less, rather than more, fair and effective.

The History and Histrionics of
Pain and Distress in Laboratory Animals

Christian E. Newcomer

Director, Division of Laboratory Animal Medicine
Professor, University of North Carolina, Chapel Hill, N.C.

I am pleased to have the opportunity to speak today. I would like to thank ILAR, for sponsoring this program, and my specialty group, the American College of Laboratory Animal Medicine, for asking me to represent some of the thinking of its constituency.

HISTORICAL PERSPECTIVE

I was concerned that there would be little left to say at the end of a day awash with the data, opinion, and histrionics that often accompany this subject. Rather than looking at the present or into the future, I thought I would begin by looking to our past in an effort to frame our progress in an historical perspective since the passage of the Animal Welfare Act (AWA) in 1966.

Congressional Remarks

I referred to the original congressional testimony on the AWA in an attempt to understand the intent of Congress regarding the alleviation of pain and distress and how these remarks translated into statutory law, regulations, and, later, policy implementation. Most of this testimony was rather uninformative and bland; however, the remarks of Congressman Claude Pepper (1965) were truly prescient. He suggested that we establish an office of Laboratory Animal Welfare in the Department of Health, Education, and Welfare; inspect and license facilities; create a process to ensure that investigators are qualified to do their work; and restrict all surgeries done for the purpose of student surgical training to terminal procedures. He stressed the importance of appropriate anesthesia, pain relief, and postoperative care.

"Three Rs"

Remarkably, the congressman discussed the "Three Rs" in his remarks. He talked about refinement and statistical analysis as a method of eliminating excessive animal use. Furthermore, he expressed the belief that institutions should provide references to research publications as part of their reporting requirement, hinting at responsibility of the scientists to demonstrate publicly the prudence and productivity of their animal research efforts. Pepper was a visionary (and/or had extraordinary staffers), and most of his ideas were adopted 20 or more years later. Is this 20-year delay evidence for the effective stonewalling of sound regulatory advances by the scientific community as the animal protection/rights organizations have claimed? Or, alternatively, should we continue to be concerned that the advocacy for animal welfare in research has been translated into an advocacy for excessive, runaway regulation as claimed by the scientific community?

The chasm between these perspectives has a pragmatic as well as a philosophical dimension:

How much effort should we invest in bridging the gap between these perspectives;

Does or should a cost benefit analysis prevail here as it does in virtually all other policy-making decisions; and most importantly,

Have we really accomplished anything to benefit the animals?

These points aside for now, Pepper and the Congress were clear in one issue: Some pain and distress may be justified and necessary. To quote Pepper, "If it is avoidable in the interest in promoting the health and protecting the lives and prolonging the lives of human beings, I am not here to oppose that."

Attention to Language

Congress was also careful in wording the 1966 AWA. Note their selection of the word "afflict" as opposed to the word "affect," with a more neutral connotation in the following statement: "The use of animals is instrumental . . . for diseases which *afflict* both humans and animals." According to Webster's dictionary, "afflict" means to distress with mental or bodily pain and to trouble greatly or grievously. Thus, Congress too has openly registered its assent that some level of pain and distress may be necessary. Later language in the AWA importunes us to "*minimize* pain and distress in animals" but not *eradicate* it. I believe these nuances in language are especially important because they are a directive from Congress to the USDA-Animal Care that a cost benefit analysis in the hands of reasonable people should prevail and that USDA-Animal Care should not attempt to embrace pure fantasy—the hyperbolic HSUS 2020 Initiative—through the development of further regulations.

Acknowledged Contributions of IACUCs

In the span of the last 20 to 30 years, the vast reduction in endemic infectious diseases in laboratory animal colonies across the United States has probably already eliminated 90 to 95% of the total pain and distress potentially experienced by ALL laboratory animals used in biomedical research. Although this process was perhaps accelerated by the advent of IACUCs in programs in which this kind of communal engagement was necessary, it was already moving forward under veterinary guidance before the 1985 AWA amendments. The research community has received few accolades for this astonishing effort. Through the IACUC in collaboration with institutional veterinarians, we now have a structure and mechanism to tackle the remaining 5 to 10% of thorny pain and distress issues and to make sound decisions about the level of attention and intervention appropriate for problem areas on a case-by-case basis.

ANTICIPATED CHANGES

Multidimensional Impact

Even without AWA regulation or policy change, we have extended our stewardship to the laboratory rat and mouse used in research by conforming to the principles of humane care and "good science." It is clear from the Animal Care Strategic Plan 2000 that the USDA-Animal Care anticipates expanding its authority in this area. This expansion could have a significant impact on the record-keeping requirements, on other dimensions of rodent care in research facilities, and on the cost of doing animal research (particularly where animal pain and distress may be involved). These potential effects give us reason to stress the need for caution and thoughtfulness in any revision of the USDA pain and distress categories or in the regulatory definition of the term "distress." In my view, the UDSA-Animal Care should rely on the NRC's *Recognition and Alleviation of Pain and Distress in Laboratory Animals* (NRC 1992) to define the term "distress" without being unduly prescriptive or trivial, and the autonomy and integrity of IACUC-scientist interactions should be preserved at the institutional level as indicated in the current AWA regulations.

Inclusion of Rodents

There are a number of other action verbs used in USDA-Animal Care Strategic Plan 2000, which indicates an aggressive agenda in the current year. I want to comment on a couple of these. One initiative involves enhancing the statutory and regulatory authority of Animal Care, presumably in reference to the inclusion of rodents in the regulations. In addition, the addition of a definition for "distress" may warrant USDA-Animal Care to identify funding for studies on distress in laboratory animals through statutory law and congressional appropriations so that scientists, IACUCs, and USDA regulators can interact on the basis of facts and not futile opinion.

Pursuit of Excellence

Another Animal Care initiative involves encouraging excellence in animal care instead of (as traditionally) encouraging compliance. While this approach is superficially appealing, its ultimate impact depends on the criteria used to define excellence and whether it imposes a larger regulatory bureaucracy on everyone to assist a few floundering organizations that fail to function at a high level without external prompting. One final point in the Animal Care Strategic Plan 2000 expresses the USDA's intention to "empower, support and develop employees." Although the concept of this intention is noble, I believe the appropriate order of events should be first to develop, then to support, and finally to empower employees. Empowered renegade employees can be a problem in any organization, but the lasting ramifications for careers of those regulated, institutional integrity, and cynicism toward regulations and regulators brings added importance to rigorous criteria for employee development in this arena. In my opinion, these criteria should be developed by expert panels that involve ACLAM board-certified veterinarians, other veterinary specialty groups, scientists with diverse expertise, and others who represent the regulated entities and the public.

RECOMMENDED ALTERNATIVE LANGUAGE

Finally, I want to offer some alternative language to improve the categorization of pain and distress in the annual reports filed by institutions concerning their use of laboratory animals. Let me preface these remarks by saying that improved categorization, per se, will contribute little or nothing to ameliorate the pain and distress experienced by laboratory animals used in scientific studies. Veterinarians and IACUCs already perform this function commendably.

Categories C and D

In my view, the language for USDA categories "C" and "D," as they are described in the USDA instructions, is adequate; and I will make the assumption that you are all familiar with that language or know where to find it. No one today has broached a sensitive subject that has been talked about ever since the act was revised in 1985-1986—the pain category designation for animals used in terminal studies. It is my belief that those animals should be in USDA Category "C" if they are anesthetized according to acceptable veterinary standards. They should not be in Category "D."

If you look back at the language of the AWA, the importance of adequate veterinary care is obvious. Congress trusted that veterinarians would provide the proper care in the mid-1960s, and that trust was extended to veterinarians and IACUCs in the 1985 AWA amendments. Congress expects that veterinarians in research will meet the standards of contemporary veterinary practice. I believe the American public widely maintains that their animals do not suffer or experience pain during the process if they die under anesthesia in the care of a veterinarian. It may have been an anesthetic or a surgical accident, grievous for both the client and the veterinarian; but it would not usually raise the specter of animal pain, distress, or suffering. I therefore believe that if an animal is anesthetized using a method that meets veterinary standards and the approval of the IACUC then this subject should not be recorded in USDA Category "D."

Category E

I propose that the reporting language could be improved by splitting Category "E." The HSUS literature appropriately alludes to the problem in Category "E" in its discussion. Animals are sometimes denied analgesics, anesthetics, and tranquilizers, not because it would affect the science, but because those substances are not clearly clinically indicated in the treatment of the kind of pain or distress that animals would suffer under those kinds of conditions. Such an example might be an animal used in an infectious disease study involving malaise, mild abdominal cramping, low-grade fever, diarrhea, and dehydration. How many and what kinds of drugs would be necessary to relieve that constellation of symptoms, and would this treatment be completely successful? It is very possible that no appropriate drugs would fall into the drug categories specified under Column "E": anesthetics, analgesics, or tranquilizers.

My solution to the language problem is minimalistic and simple: split USDA Category "E" into two categories, "E1" and "E2," and retain the original language for "E1." For Category E2, fashion some language to express our need to include other types of vague pain, distress, and suffering, which are poorly characterized and for which the anesthetics, analgesics, sedatives, and tranquilizers may not be deemed appropriate. I offer the following language for the new Category "E2": "Provide the common names and approximate numbers of animals used in applications that resulted in significant pain or distress but for which palliative efforts either were not undertaken or were deemed ineffective in the opinion of the IACUC and the attending veterinarian due to the type of pain or distress syndrome involved and the current status of applicable pharmacological and non-pharmacological therapeutic options."

The language recommendations above represent my attempt to create a catchall, to put the discrimination for inclusion in this catchall with the IACUC and the attending veterinarian, where it truly belongs. I do not believe that an exhaustive revision of the current reporting requirement coupled with annual USDA-Animal Care inspections and protocol reviews will improve the ability of institutions to conduct science under humane conditions. Congress stood behind the IACUC in its statutory language, which reflects their belief that institutions are competent to proceed the right way. I believe the good faith principle should be extended to IACUCs and veterinarians because evidence exists that we have performed very well and have made significant progress.

CONCLUSION

I have always believed that complicated policy matters are best informed by the simplest and clearest expression of principle. I want to leave you with a guiding principle that is taken from *The Little Prince,* by Antoine de Saint-Exupéry (1943) during an exchange between the Little Prince and the king:

> "If I order the general to fly from one flower to another like a butterfly, or to write a tragic drama, or to change himself into a seabird, and if the general did not carry out the order that he had received, which one of us would be in the wrong," the king demanded, "the general or myself?"

> "You," said the little prince firmly.

> "Exactly. One must require from each one the duty which one can perform," the king went on. "Accepted authority rests first of all on reason." (p. 38)

REFERENCES

de Saint-Exupéry A. 1943. The Little Prince. San Diego: Harcourt Brace.
NRC [National Research Council]. 1992. Recognition and Alleviation of Pain and Distress in Laboratory Animals. Washington, D.C.: National Academy Press.
Pepper C. Congressional Committee Report, September 30, 1965. Washington, D.C.: GPO.

Panel Discussion with All Speakers

DR. DELL: We have heard a great deal of information today, and I am sure that people have many questions. I invite the speakers and other people in the audience to ask questions and share additional information that would be useful.

DR. DE HAVEN: I would be remiss if I did not say that this workshop has been hugely helpful. Although you would expect me to make that statement, I genuinely believe we have heard a very broad spectrum of perspectives and ideas, and I think that you have given us much to think about. I particularly appreciate the fact that ILAR and NIH have hosted this forum. You understand that there is a law called the Federal Advisory Committee Act that would keep us from simply assembling without going through a very complicated and due process and enlisting a group of experts to help us draft a policy, a regulation, or even guidelines. However, nothing keeps anyone from making suggestions to us, and we can assure you that we will give all suggestions due consideration, whether they are coming from an individual, an organization, or an association; from industry, the animal protection community, or simply the general public at large. In spite of those limitations, we greatly appreciate these kinds of public meetings (whether we or someone else hosts them), and we certainly look forward to your comments through the advanced notice of proposed rule making, which we intend to publish.

After we publish a notice in the Federal Register and solicit your comments on that advance notice of proposed rule making, the end result could be that we will decide to do nothing, or that we will indeed decide to initiate some kind of rulemaking process. If so, we will then publish a proposed rule and provide an opportunity for comment.

Let me return to the advance notice of proposed rulemaking, which we hope to have published in the Federal Register in the next month or less. One of the ideas we are considering and on which we would like you to comment is a proposed definition of distress. We will suggest various pain and distress categorization schemes. We would certainly like your comments on anything we include in that Federal Register notice, but at the same time, please do not feel limited in giving us your comments regarding an example we might publish.

In the interim, we indeed will pursue a change to Policy 11. Because any rulemaking effort we might undertake is at best a 3-year process, I think we owe it to you and to our inspectors to establish a working definition of distress if we are going to require the same kind of considerations with distressful procedures as we have in the past with painful ones. Such an interim policy could be changed through a regulatory effort to become a much more enforceable definition or one that would carry more weight than a policy. Of course, one of the benefits of the definition or anything we put into policy is that it is much easier to change.

MS. LISS (Cathy Liss, Animal Welfare Institute): I have a question regarding remarks made by Dr. Harkness, who I believe has left. He mentioned a protocol that was put before an IACUC that called for two major operative procedures for cats. I wondered how that situation was resolved.

First, my hat is off to the IACUC for having done the right thing in admonishing the researcher about having two major procedures. Apparently they discussed it extensively and resolved it to prevent a recurrence. Second, Dr. Harkness mentioned that NIH had provided funding in the meantime, so, I am wondering where it went from there.

DR. GARNETT (Nelson Garnett, Office of Laboratory Animal Welfare, NIH): That description shocked me as much as everyone else. Under NIH grants review policy, it is actually not allowed for the study section even to review the application without evidence of protocol review. That situation may very well be the subject of an investigation.

DR. NEWCOMER (Christian Newcomer, University of North Carolina): At the University of North Carolina, we have a large number of hemophilic dogs and pigs. We currently do not record these animals in Category E although, of course, we do think that when they have bleeding episodes (which are frequent in those animals), they do experience distress despite the fact that they are given analgesics in response to visibly painful conditions, joint bleeds, and so forth. If you do revise Policy 11 with an acceptable definition of distress, it is very likely that these animals would clearly fit that definition of distress, and there still would be no place to record those animals on the annual report other than in Category C and/or E depending on their particular situation. This then is an example of where, despite what you do for the animal, you recognize that the animal has distress when you have finished administering therapies. It appears that there is no appropriate place to include that information in the form. Furthermore, it appears that it would require only a brief explanation because bleeding is intrinsic to their disease. I am interested to hear how you think that kind of event would be properly recorded in the future and in the present.

DR. DE HAVEN: You may be making a good case, Dr. Newcomer, for a reason we need to modify the existing categories because these animals do appear to fit currently in Column E. More importantly, and I think you and others have made the point throughout the day, that many people (not the least of whom is our inspector) must exercise good, professional judgment.

Whether we like it or not, the role of the inspector is to monitor the exercise of professional judgment at the institution (whether by the IACUC, the investigator or indeed by the attending veterinarian) and clearly to give the benefit of the doubt to the institution unless it is clearly departing from established policy and then indeed, as distasteful as it may be, correct the institution. That correction may be in the form of nothing more than a discussion at the time of the inspection; it may generate some correspondence after the inspection; it could in fact be a citation on the inspection report depending on the situation.

I would hesitate to make policy on hemophilic drugs in pigs from this podium without knowing all of the facts and so I will not do so. Nevertheless, it is critical to compose a definition of distress, and then we would hope that at your institution and everyone else's there would be good use of professional judgment regarding which category is selected among the categories we have in place at the time.

DR. DELL: I would like to follow up on Dr. Newcomer's question. We heard from Dr. Rich, who is just one of many who have recently shared information with ILAR as we try to project the future, changing use of animals and identify the resulting cost implications. We have heard from a great number of people that because of our ability to manipulate the mouse genome, we will actually be able to actually create real models of real human disease and then try to understand the complex molecular events, identify some point where one could develop a small molecule that would interrupt the pathway, and ultimately develop a drug that could be used in humans to treat disease. Particularly in mice, it is anticipated that there will be a huge number of human disease models, many of which will be associated with episodes or continuous pain. For example, arthritis has a painful component to it. In the hemophilia model, the animals are fine most of the time. They have trouble only during acute episodes. How do you try to categorize or deal with models like that?

DR. DE HAVEN: I would ask the same question. The whole area of transgenics, as Dr. Rich has indicated, will change the norm in the next 20 (or probably fewer) years. I think we are just beginning to deal with those issues, and perhaps we need to set aside a whole separate process to look at the area of transgenics. In the meantime, I think we are starting to deal with it when we talk about distress for example. Distress would include such things as induced disease processes that cause the animal to suffer the consequences, feel uncomfortable, feel discomfort, or whatever condition it might be. Whether a disease process is induced because of genetic manipulation or because of infecting an animal with an infectious agent, I think we should ultimately question the effect on the animal—not so much the process, again as Dr. Rich said, but the end result, the outcome for the animal. If the animal experiences pain and/or distress, then it needs to be put into the appropriate category, whatever category we have in place at the time.

DR. HAMPSHIRE (Victoria Hampshire, Advanced Veterinary Applications): Many people who breed transgenics actually breed heterozygotes so that the heterozygote is not typically sick; they cross two heterozygotes. There are ways of managing transgenic populations to offset or mitigate distress, although people without previous experience in managing distress would benefit from the exemplars Dr. Rich described.

You could perhaps require a pilot of a small number of breeding pairs, with the retrospective classification back from the IACUC, to focus more on the wording. Careful wording will provide both the efficiency the scientist wants and the answers the IACUC wants. Then you are reporting much more accurately.

DR. RICH: Basically, I do not disagree with what you are saying, although I would point out that the majority of diseases likely to be created will be created not by transgenic manipulation but instead by gene knockout manipulation. In most cases, heterozygotes will not have the phenotype because most of the diseases will be complex, just as most diseases we see are not autosomatognotic. Most will be complex diseases involving multiple interactions between genes and often requiring double hits. The issue is really very complicated.

The best thing we can do in terms of reducing animals in the process of the extraordinarily expensive as well as lengthy process of breeding transgenic and knockout animals is to encourage the development of embryo freezing so that we can maintain the lowest possible number of necessary colonies. We grow up the animals as we are ready to study them, as opposed to breeding up large numbers of animals and not being able to study them in a timely way. Whenever we develop substantial transgenic and knockout facilities, we should also develop embryo freezing programs.

MS. WOLFE (Lisa Wolfe, People for the Ethical Treatment of Animals): To seek knowledge, to cure, and to prevent disease is noble; but to seek knowledge at any cost is not. To seek knowledge by experimenting on animals or any other involuntary participant is to do so at great cost. That cost is the dulling of our society's values of nonviolence, justice, compassion, and empathy.

No one has the right to exploit another being for one's own benefit without his or her consent. If our goal is truly to minimize pain and distress in animals used in experiments, the most effective way, and indeed the only way, is to recognize that we do not have the right to use them.

DR. BENNETT (Taylor Bennett, University of Illinois at Chicago): This has been a very informative day. We have been discussing a very important issue, which I believe calls for a more formal regulatory process, which publicizes the issues. Through that process, people have the chance to think through and discuss the issues and then end up with something that most of us can accept. I think that what we have done here today is very important; however, I think we need more time and advance notice to consider and respond to this and other new policies in the future. More input from a group such as our laboratory animal community will result in more "buy-in" when policies are in place.

Appendix A

APHIS/USDA Policy 11 and Policy 12

POLICY #11—PAINFUL/DISTRESSFUL PROCEDURES—APRIL 14, 1997

References: AWA Sections 13(a)(3), 13(a)(7), 13(e)(2, 3) and 9 CFR, Part 2, Sections 2.31(d)(1)(i,ii,iii,iv), 2.31(e)(4), 2.33(b)(4) and 9 CFR, Part 3, Section 3.6(b)(5,6,7)

History: Replaces letters dated May 8, 1992, November 7, 1991, November 9, 1990, and March 1, 1990.

Justification: Provides requested guidance. Procedures involving animals will avoid or minimize discomfort, distress and/or pain.

Policy:

A painful procedure is defined as any procedure that would reasonably be expected to cause more than slight or momentary pain and/or distress in a human being to which that procedure is applied. The Institutional Animal Care and Use Committee (IACUC) is responsible for ensuring that investigators have appropriately considered alternatives to any procedures that may cause more than slight or momentary pain or distress. A written narrative description of the methods and sources used to search for alternatives must be provided. Where specific testing procedures are required by Federal law, the CFR references or other legal guidelines requiring them should be noted.

Examples of procedures that can be expected to cause more than momentary or slight pain include, but are not limited to, the following:

- **Terminal Surgery** is considered a painful procedure which is alleviated by anesthesia.
- **Freund's Complete Adjuvant** used for antibody production may cause results ranging from momentary or slight pain to severe pain depending on the product, procedure, and species.
- **Ocular and Skin Irritancy Testing.** The dosing procedure itself is generally not painful but the reaction caused by the product being tested may cause pain.

Examples of procedures that may cause more than momentary or slight distress include, but are not limited to, the following:

- **Food or water deprivation** beyond that necessary for normal presurgical preparation.
- **Noxious electrical shock** that is not immediately escapable.
- **Paralysis or immobility** in a conscious animal.

Many procedures, including any of those in the lists above, may cause both pain and distress. An example of a procedure that can be expected to cause more than momentary or slight pain as well as distress would be a study involving extensive irradiation.

Animals exhibiting signs of pain, discomfort, or distress such as decreased appetite/activity level, adverse reactions to touching inoculated areas, open sores/ necrotic skin lesions, abscesses, lameness, conjunctivitis, corneal edema, and photophobia are expected to receive appropriate relief unless written scientific justification is provided in the animal activity proposal and approved by the IACUC.

Research facilities must have a mechanism in place for ensuring that animals are reported in the appropriate pain category on the annual report (APHIS Form 7023). Individual animals that do not experience pain/distress from testing procedures should be reported in column C. Individual animals experiencing pain/ distress which is alleviated with anesthetics, analgesics, sedatives and/or tranquilizers should be reported in column D. This category includes terminal surgery under anesthesia. Individual animals in which needed anesthetics, analgesics, sedatives, and/or tranquilizers are withheld should be reported in column E. For all column E animals, a written justification, approved by the IACUC, must be provided, including CFR references or other guidelines if appropriate.

POLICY #12 — CONSIDERATION OF ALTERNATIVES TO PAINFUL/DISTRESSFUL PROCEDURES — JUNE 21, 2000

References: AWA Section 13(a)(3)(B), 9 CFR, Part 2, Section 2.31 (d)(1)(ii)and (e), 9 CFR, Part 2, Section 2.32 (c)(2) and (5)(ii), Animal Welfare Information Center

History: Provides guidance on the requirement to provide a written narrative of the consideration of alternatives to painful and distressful procedures. Replaces Policy #12 dated April 14, 1997.

Justification: The Animal Welfare Act (AWA) regulations require principal investigators to consider alternatives to procedures that may cause more than momentary or slight pain or distress to the animals and provide a written narrative of the methods used and sources consulted to determine the availability of alternatives, including refinements, reductions, and replacements.

Policy:

Alternatives or alternative methods are generally regarded as those that incorporate some aspect of replacement, reduction, or refinement of animal use in pursuit of the minimization of animal pain and distress consistent with the goals of the research. These include methods that use non-animal systems or less sentient animal species to partially or fully replace animals (for example, the use of an in vitro or insect model to replace a mammalian model), methods that reduce the number of animals to the minimum required to obtain scientifically valid data, and methods that refine animal use by lessening or eliminating pain or distress and, thereby, enhancing animal well-being. Potential alternatives that do not allow the attainment of the goals of the research are not, by definition, alternatives.

A fundamental goal of the AWA and the accompanying regulations is the minimization of animal pain and distress via the consideration of alternatives and alternative methods. Toward this end, the regulations state that any proposed animal activity, or significant changes to an ongoing animal activity, must include:

1. a rationale for involving animals, the appropriateness of the species, and the number of animals to be used;
2. a description of procedures or methods designed to assure that discomfort and pain to animals will be limited to that which is unavoidable in the conduct of scientifically valuable research, and that analgesic, anesthetic, and tranquilizing drugs will be used where indicated and appropriate to minimize discomfort and pain to animals;
3. a written narrative description of the methods and sources used to consider

alternatives to procedures that may cause more than momentary or slight pain or distress to the animals; and

4. the written assurance that the activities do not unnecessarily duplicate previous experiments.

We believe that the performance of a database search remains the most effective and efficient method for demonstrating compliance with the requirement to consider alternatives to painful/distressful procedures. However, in some circumstances (as in highly specialized fields of study), conferences, colloquia, subject expert consultants, or other sources may provide relevant and up-to-date information regarding alternatives in lieu of, or in addition to, a database search. When other sources are the primary means of considering alternatives, the Institutional Animal Care and Use Committee (IACUC) and the inspecting Veterinary Medical Officer should closely scrutinize the results. Sufficient documentation, such as the consultant's name and qualifications and the date and content of the consult, should be provided to the IACUC to demonstrate the expert's knowledge of the availability of alternatives in the specific field of study. For example, an immunologist cited as a subject expert may or may not possess expertise concerning alternatives to in vivo antibody production.

When a database search is the primary means of meeting this requirement, the narrative must, at a minimum, include:

1. the names of the databases searched;
2. the date the search was performed;
3. the period covered by the search; and
4. the key words and/or the search strategy used.

The Animal Welfare Information Center (AWIC) is an information service of the National Agricultural Library specifically established to provide information about alternatives. AWIC offers expertise in formulation of the search strategy and selection of key words and databases, access to unique databases, on- and off-site training of institute personnel in conducting effective alternatives searches, and is able to perform no-cost or low-cost electronic database searches. AWIC can be contacted at (301) 504-6212, via E-mail at awic@nal.usda.gov, or via its Web site at http://www.nal.usda.gov/awic. Other excellent resources for assistance with alternative searches are available and may be equally acceptable.

Regardless of the alternatives sources(s) used, the written narrative should include adequate information for the IACUC to assess that a reasonable and good faith effort was made to determine the availability of alternatives or alternative methods. If a database search or other source identifies a bona fide alternative method (one that could be used to accomplish the goals of the animal use proposal), the written narrative should justify why this alternative was not used.

The written narrative for federally mandated animal testing (for example, testing product safety/efficacy/potency) need only to include a citation of the appropriate government agency's regulation and guidance documents. Mandating agency guidelines should be consulted since they may provide alternatives (for example, refinements such as humane endpoints or replacements such as the Murine Local Lymph Node Assay) that are not included in the Code of Federal Regulations. If a mandating agency-accepted alternative is not used, the principal investigator should explain the reason in the written narrative.

Alternatives should be considered in the planning phase of the animal use proposal. When a proposal is modified during its performance, significant changes are subject to prior review by the IACUC, including the review of the implications of those changes concerning the availability of alternatives. Although additional attempts to identify alternatives or alternative methods are not required by Animal Care at the time of each annual review of the animal protocol, Animal Care would normally expect the principal investigator to reconsider alternatives at least once every 3 years, consistent with the triennial review requirements of the Public Health Service Policy (IV,C,5).

Appendix 2

Proposed Rulemaking

[Federal Register: July 10, 2000 (Volume 65, Number 132)]
[Proposed Rules]
[Page 42304-42305]
From the Federal Register Online via GPO Access [wais.access.gpo.gov]
[DOCID:fr10jy00-18]

Proposed Rules
Federal Register

This section of the FEDERAL REGISTER contains notices to the public of the proposed issuance of rules and regulations. The purpose of these notices is to give interested persons an opportunity to participate in the rule making prior to the adoption of the final rules.

[[Page 42304]]

DEPARTMENT OF AGRICULTURE

Animal and Plant Health Inspection Service

9 CFR Parts 1 and 2

[Docket No. 00-005-1]

Animal Welfare; Definitions for and Reporting of Pain and Distress

AGENCY: Animal and Plant Health Inspection Service, USDA.

ACTION: Request for comments.

SUMMARY: We are considering several changes to the Animal Welfare regulations to promote the humane treatment of live animals used in research, testing, and teaching and to improve the quality of information we report to Congress concerning animal pain and distress. Specifically, we are considering adding a definition for the term "distress." Although this term is used throughout the Animal Welfare regulations, it is not defined. The addition of such a definition would clarify what we consider to be "distress" and could help assist research facilities to recognize and minimize distress in animals in accordance with the Animal Welfare Act (AWA).

We are also considering replacing or modifying the system we use to classify animal pain and distress. Professional standards regarding the recognition and relief of animal pain and distress have changed significantly since we established our classification system. Some biomedical research professionals and animal welfare advocates believe our classification system is outdated and inadequate. A different categorization system could produce data that more accurately depict the nature of animal pain or distress and provide a better tool to measure efforts made to minimize animal pain and distress at research facilities.

We are soliciting public comments on the changes we are considering. We are also interested in obtaining information on specific pain and distress classification systems other than the one we now use.

DATES: We invite you to comment on this docket. We will consider all comments that we receive by September 8, 2000.

ADDRESSES: Please send your comment and three copies to: Docket No. 00-005-1, Regulatory Analysis and Development, PPD, APHIS, Suite 3C03, 4700 River Road, Unit 118, Riverdale, MD 20737-1238.

Please state that your comment refers to Docket No. 00-005-1. You may read any comments that we receive on this docket in our reading room. The reading room is located in room 1141 of the USDA South Building, 14th Street and Independence Avenue, SW., Washington, DC. Normal reading room hours are 8 a.m. to 4:30 p.m., Monday through Friday, except holidays. To be sure someone is there to help you, please call (202) 690-2817 before coming.

APHIS documents published in the Federal Register, and related information, including the names of organizations and individuals who have commented on APHIS dockets, are available on the Internet at http://www.aphis.usda.gov/ppd/rad/webrepor.html.

FOR FURTHER INFORMATION CONTACT: Dr. Jodie Kulpa, Staff Veterinarian, AC, APHIS, 4700 River Road Unit 84, Riverdale, MD 20737-1234; (301) 734-7833.

SUPPLEMENTARY INFORMATION:

Background

Under the Animal Welfare Act (AWA) (7 U.S.C. 2131 et seq.), the Secretary of Agriculture is authorized to promulgate standards and other requirements regarding the humane handling, care, treatment, and transportation of certain animals by dealers, research facilities, exhibitors, carriers and intermediate handlers. The Secretary has delegated responsibility for administering the AWA to the Animal and Plant Health Inspection Service (APHIS) of the United States Department of Agriculture (USDA). Regulations established under the AWA are contained in the Code of Federal Regulations (CFR) in title 9, parts 1, 2, and 3 (referred to below as the regulations). Part 1 contains definitions for terms used in parts 2 and 3. Part 2 contains general requirements for regulated parties. Part 3 contains specific requirements for the care and handling of certain animals.

We are soliciting comments on an approach, discussed below, for amending the regulations by defining "distress" in part 1 and by modifying or replacing the animal pain and distress classification system in part 2.

Definition for Distress

In the regulations, we define a "painful procedure" as any procedure that would reasonably be expected to cause more than slight or momentary pain or distress in a human being to which that procedure was applied. Although we use the term "distress" in this definition and elsewhere in the regulations, there is no definition for distress in the regulations. We are considering adding such a definition because of requests from the biomedical research community and animal advocacy groups. These parties have asked USDA to provide guidance on what is considered to be distress in a procedure involving research animals in order to improve recognition of animal distress, to classify and report it more accurately, and to create a heightened awareness of the regulations' requirement to minimize animal distress and pain.

Pain and Distress Classification System

Section 13(a)(7)(B) of the AWA requires research facilities to annually provide "information on procedures likely to produce pain or distress in any animal." In accordance with the AWA, the regulations at Sec. 2.36 require facilities that use or intend to use live animals for research, tests, experiments, or teaching to submit an annual report to the Animal Care Regional Director for the State where the facility is located. Among other things, the report must state the common names and the numbers of animals upon which teaching, experiments, research, surgery, or tests were conducted involving: (1) No pain, distress, or use of pain-relieving drugs; (2) accompanying pain or distress to the animals and for which appropriate anesthetic, analgesic, or tranquilizing drugs were used; and (3) accompanying pain or distress to the animals and for which the use of appropriate anesthetic, analgesic, or tranquilizing drugs would have adversely affected the procedures, results, or interpretation of the teaching, research, experiments, surgery, or tests.

[[Page 42305]]

To provide these data, each research facility must assess the potential for animal pain or distress associated with the proposed procedures. This assessment is performed prospectively (i.e., before the procedure) and typically forms the basis for the pain and distress report provided by the facility to USDA. The assessment, therefore, is an estimate based on professional judgment, knowledge, and experience, and the resulting report may or may not accurately reflect the conditions the animals actually experience. The research facility can, as an option, retrospectively (i.e., during or after the procedure) assess the animal pain and distress observed and report these results. We do not know how often facilities perform retrospective reporting.

There is no provision in the current classification system to address some areas identified by the research community and animal advocacy groups. For example, the current system does not include a means to report:

An assessment of the relative intensity or duration of pain or distress either observed in the animal or anticipated to be experienced by the animal;

An assessment of the anticipated or observed efficacy of the pain- or distress-relieving agent provided to animals undergoing a painful or distressful procedure;

A distinction between procedures causing animal pain and procedures causing animal distress;

Animals that were prevented from experiencing pain or distress by the appropriate and effective use of pain- or distress-relieving methods or procedures (e.g., well-anesthetized animals that undergo terminal surgery);

Animals that did not experience pain or distress due to the appropriate and effective use of pain- or distress-relieving methods or procedures other than anesthetic, analgesic, or tranquilizing agents;

Animals that experience unrelieved pain or distress for a reason other than that the use of anesthetic, analgesic, or tranquilizing drugs would have adversely affected the procedures, results, experiments, surgery, or tests; or

Animals that experience pain or distress without having been used in a procedure (e.g., illness in animals that have been genetically altered to develop disease).

We are aware of several alternative pain and distress classification systems. For example, the system adopted by the Canadian Council on Animal Care, "Categories of Invasiveness in Animal Experiments," may be viewed on the Internet at http://www.ccac.ca/english/categ.htm. The system proposed by the Humane Society of the United States may be viewed on the Internet at http://hsus.org/ programs/research/usda_proposed_scale.html.\1\ Other classification systems, varying greatly in complexity, are in use in other countries, such as Switzerland and Sweden.

\1\ If you do not have access to the Internet, you may obtain a copy of the system adopted by Canadian Council on Animal Care or the system proposed by the Humane Society of the United States by contacting the person listed under FOR FURTHER INFORMATION CONTACT at the beginning of this document.

Modifying the current USDA system, in lieu of replacing it, could also be an option. This could involve replacing or redefining the existing categories to:
Separately report pain and distress;
Quantify pain and distress intensity and duration;
Separately classify anesthetized or otherwise treated animals undergoing potentially painful procedures but not experiencing pain or distress; or
Modify the system in other ways.

We invite your comments on adding a definition for distress to the regulations and replacing or modifying our animal pain and distress classification system. We are particularly interested in soliciting comments addressing the following questions:

1. Would adding a definition for distress to the regulations help institutions using animals for research, testing, or teaching better recognize, minimize, and report animal distress?

2. If a definition for distress is added to the regulations, what key elements should be included in that definition?

3. What are the benefits and limitations of our pain and distress classification system?

4. Should our animal pain and distress classification system be modified or replaced? If so, what specific modifications or alternate classification systems should we consider?

5. Should animal pain and distress be prospectively or retrospectively reported?

Written comments should be submitted within the 60-day comment period specified in this document (see DATES and ADDRESSES).

Executive Order 12866

This action has been reviewed under Executive Order 12866. The action has been determined to be not significant for the purposes of Executive Order 12866 and, therefore, has not been reviewed by the Office of Management and Budget.

Authority: 7 U.S.C. 2131-2159; 7 CFR 2.22, 2.80, and 371.2(g).

Done in Washington, DC, this 3rd day of July 2000.
Bobby R. Acord,
Acting Administrator, Animal and Plant Health Inspection Service.
[FR Doc. 00-17280 Filed 7-7-00; 8:45 am]
BILLING CODE 3410-34-P

[Federal Register: August 21, 2000 (Volume 65, Number 162)]
[Proposed Rules
[Page 50667]
From the Federal Register Online via GPO Access [wais.access.gpo.gov]
[DOCID:fr21au00-28]

DEPARTMENT OF AGRICULTURE

Animal and Plant Health Inspection Service

9 CFR Parts 1 and 2

[Docket No. 00-005-2]

Animal Welfare; Definitions for and Reporting of Pain and Distress

AGENCY: Animal and Plant Health Inspection Service, USDA.

ACTION: Notice of extension of comment period.

SUMMARY: We are extending the comment period for our request for comments concerning several changes we are considering making to the Animal
Welfare regulations to promote the humane treatment of live animals used in
research, testing, and teaching and to improve the quality of information we report
to Congress concerning animal pain and distress. This action will allow interested
persons additional time to prepare and submit comments.

DATES: We invite you to comment on Docket No. 00-005-1. We will consider
all comments that we receive by November 7, 2000.

ADDRESSES: Please send your comment and three copies to: Docket No.
00-005-1, Regulatory Analysis and Development, PPD, APHIS, Suite 3C03, 4700
River Road, Unit 118, Riverdale, MD 20737-1238.
    Please state that your comment refers to Docket No. 00-005-1.
    You may read any comments that we receive on this docket in our reading
room. The reading room is located in room 1141 of the USDA South Building,
14th Street and Independence Avenue, SW., Washington, DC. Normal reading
room hours are 8 a.m. to 4:30 p.m., Monday through Friday, except holidays. To
be sure someone is there to help you, please call (202) 690-2817 before coming.
    APHIS documents published in the Federal Register, and related informa-

tion, including the names of organizations and individuals who have commented on APHIS dockets, are available on the Internet at http://www.aphis.usda.gov/ppd/rad/webrepor.html.

FOR FURTHER INFORMATION CONTACT: Dr. Jodie Kulpa, Staff Veterinarian, AC, APHIS, 4700 River Road Unit 84, Riverdale, MD 20737-1234; (301) 734-7833.

SUPPLEMENTARY INFORMATION:

Background

On July 10, 2000, we published in the Federal Register (65 FR 42304-42305, Docket No. 00-005-1) a request for comments on several changes we are considering making to the Animal Welfare regulations to promote the humane treatment of live animals used in research, testing, and teaching and to improve the quality of information we report to Congress concerning animal pain and distress. Specifically, we are considering adding a definition for the term "distress" and replacing or modifying the system we use to classify animal pain and distress.

Comments in response to our request for comments were required to be received on or before September 8, 2000. In response to requests from the public, we are extending the comment period on Docket No. 00-005-1 for an additional 60 days. This action will allow interested persons additional time to prepare and submit comments.

Authority: 7 U.S.C. 2131-2159; 7 CFR 2.22, 2.80, and 371.7.

Done in Washington, DC, this 15th day of August 2000.
Bobby R. Acord,
Acting Administrator, Animal and Plant Health Inspection Service.
[FR Doc. 00-21173 Filed 8-18-00; 8:45 am]
BILLING CODE 3410-34-P

Appendix C

Glossary of Abbreviations

AAALAC International, Association for Assessment and Accreditation of Laboratory Animal Care

AALAS, American Association for Laboratory Animal Science

ACLAM, American College of Laboratory Animal Medicine

AVMA, American Veterinary Medical Association

AWA, Animal Welfare Act

FASEB, Federation of American Societies for Experimental Biology

HSUS, Humane Society of the United States

IACUC, institutional animal care and use committee

IASP, International Association for the Study of Pain

ILAR, Institute for Laboratory Animal Research

NABR, National Association for Biomedical Research

NIH, National Institutes of Health

NRC, National Research Council

OLAW, Office of Laboratory Animal Welfare

PHS, Public Health Service

PRIM&R, Public Responsibility in Medicine & Research

SCAW, Scientists Center for Animal Welfare

USDA, United States Department of Agriculture

Appendix D

Meeting Participants

Lynn C. Anderson, DVM, Senior Director, Comparative Medicine/LAR, Merck Research Laboratories, Rahway, N.J.

Kathryn Bayne, MS, PhD, DVM, Associate Director, American Association for the Accreditation, of Laboratory Animal Care International, Rockville, Md.

B. Taylor Bennett, DVM, PhD, Biological Resources Laboratory, University of Illinois, Chicago, Ill.

Marcelo Couto, DVM, PhD, Scientific Advisory Committee, American Association for Laboratory Animal Science, Memphis, Tenn.

W. Ron DeHaven, DVM, Deputy Administrator, U.S. Department of Agriculture, Animal and Plant Health Inspection Service, Animal Care, Riverdale, Md.

Nelson Garnett, DVM, Office of Laboratory Animal Welfare, National Institutes of Health, Bethesda, Md.

G. F. Gebhart, PhD, Professor and Head, Department of Pharmacology, College of Medicine, University of Iowa, Iowa City, Iowa

Victoria Hampshire, VMD, Advanced Veterinary Applications, Bethesda, Md.

John E. Harkness, DVM, Laboratory Animal Veterinarian, College of Veterinary Medicine, Mississippi State University, Mississippi State, Miss.

J. R. Haywood, PhD, Department of Pharmacology, University of Texas Health Science Center, San Antonio, Tex.

Alicia Karas, DVM, MS, Associate Professor of Anesthesiology, Department of Clinical Sciences, Tufts University School of Veterinary Medicine, North Grafton, Mass.

Christian E. Newcomer, VMD, Director, Division of Laboratory Animal Medicine, University of North Carolina, Chapel Hill, N.C.

Robert R. Rich, MD, Executive Associate Dean and Professor of Medicine and Microbiology and Immunology, Emory University School of Medicine, Atlanta, Ga.

Andrew N. Rowan, DPhil, Senior Vice President of Research, Education, and International Issues, Humane Society of the United States, Washington, D.C.

Appendix E

Meeting Agenda

**WORKSHOP ON DEFINITION OF PAIN AND DISTRESS AND
REPORTING REQUIREMENTS**

Meeting Room: Auditorium
June 22, 2000
AGENDA

Speakers will provide operative definitions of pain and distress in laboratory animals and discuss whether the pain and distress classifications used in the USDA annual report should be modified. The impact of the current reporting system and any proposed reporting system on the regulated entities will also be assessed.

8:30 – 8:45	**Ralph B. Dell, MD** Director, ILAR
8:45 – 9:00	**W. Ron DeHaven, DVM** Deputy Administrator Animal Care USDA/APHIS
9:00 – 9:15	**Nelson Garnett, DVM** Director, Office of Laboratory Animal Welfare National Institutes of Health

9:15 – 9:45   **Kathryn Bayne, MS, DVM, PhD**
                       Associate Director
                       American Association for the Accreditation of Laboratory
                       Animal Care International

9:45 – 10:15   **G. F. Gebhart, PhD**
                       Head, Department of Pharmacology
                       University of Iowa

10:15 – 10:30   Break

10:30 – 11:00   **Andrew Rowan, DPhil**
                       Senior Vice President for Research, Education, and
                       International Issues
                       Humane Society of the United States

11:00 – 11:30   **Alicia Karas, DVM, MS**
                       Associate Professor of Anesthesiology, Department of
                       Clinical Sciences
                       Tufts University School of Veterinary Medicine

11:30 – 12:00 PM   **B. Taylor Bennett, DVM, PhD**
                       Biological Resources Laboratory
                       University of Illinois

12:00 – 12:45   Lunch

12:45 – 1:15   **Marcelo Couto, DVM, PhD**
                       Scientific Advisory Committee
                       American Association for Laboratory Animal Science

1:15 – 1:45   **J.R. Haywood, PhD**
                       Department of Pharmacology
                       University of Texas Health Science Center

1:45 – 2:15   **Lynn C. Anderson, DVM**
                       Senior Director, Comparative Medicine/LAR
                       Merck Research Laboratories

2:15 – 2:45   **John Harkness, DVM**
                       Laboratory Animal Veterinarian
                       Mississippi State University College of Veterinary Medicine

2:45 – 3:00   Break

3:00 – 3:30   **Victoria Hampshire, VMD**
                   Advanced Veterinary Applications

3:30 – 4:00   **Robert Rich, MD**
                   Emory University School of Medicine

4:00 – 4:30   **Christian Newcomer, VMD**
                   Director, Division of Laboratory Animal Medicine
                   University of North Carolina

4:30 – 5:00   Panel Discussion with All Speakers

5:00   Adjourn

Appendix F

Biographical Sketches of Committee Members

Joanne Zurlo *(Chair)* is Associate Director, Center for Alternatives to Animal Testing, Johns Hopkins School of Hygiene and Public Health. Her research interests have been in the areas of chemical carcinogenesis, molecular biology, and in vitro toxicology. She has held faculty positions at Dartmouth Medical School in the Department of Pharmacology and Toxicology and at Johns Hopkins School of Hygiene and Public Health in the Department of Environmental Health Sciences.

Adele Douglass is Director of Public Policy for the American Humane Association. She is experienced in public policy concerning the use of animals in research and in agriculture. She has a broad range of knowledge of animal welfare issues, including agriculture and biomedical research uses of animals and is a recognized expert on animal welfare and the public's perception of those issues.

Randall J. Nelson is a Professor in the Department of Anatomy and Neurobiology, College of Medicine, University of Tennessee. He is a neuroscientist and an expert in the use of nonhuman primates. His studies concern the role of the somatosensory cortex in receiving peripheral sensory information, integrating it with various central inputs, and contributing to the control of movement.

William S. Stokes is Associate Director for Animal and Alternative Resources, Environmental Toxicology Program, NIEHS and co-chair of the Interagency Coordinating Committee on Validation of Alternative Methods. His research interests are toxicological methods, including development, validation, and acceptance of new animal models, and improved toxicological test systems.

Jerrold Tannenbaum is Professor of Population Health and Reproduction, School of Veterinary Medicine, University of California at Davis. He did graduate work in philosophy at Rockefeller University and Cornell University and obtained his JD from Harvard Law School. He is the author of numerous papers on veterinary and animal law and ethics, has written *Veterinary Ethics,* the first and only comprehensive book on veterinary ethics, and has spoken on ethical and legal issues relating to animals to a variety of audiences ranging from veterinary students to humane societies.